LORDS OF THE FLY

Robert E. Kohler

LORDS OF THE FLY

Drosophila Genetics
and the
Experimental Life

The University of Chicago Press
Chicago & London

Robert Kohler is professor in the Department of History and Sociology of Science at the University of Pennsylvania and author of *Partners in Science,* also published by the University of Chicago Press.

The University of Chicago Press, Chicago 60637
The University of Chicago Press, London
© 1994 by The University of Chicago
All rights reserved. Published 1994
Printed in the United States of America
04 03 02 01 00 99 98 97 96 95 94 5 4 3 2 1
ISBN: 0–226–45062–7 (cloth)
0–226–45063–5 (paper)

Library of Congress Cataloging-in-Publication Data
Kohler, Robert E.
    Lords of the fly : Drosophila genetics and the experimental life /
Robert E. Kohler.
        p.   cm.
    Includes bibliographical references (p.   ) and index.
        1. Research—Methodology.   2. Scientists.   3. Drosophila—
Research.   I. Title.
Q180.55.M4K63   1994
502.8—dc20   93-29436

Contents

Illustrations

For Jack and Willie

and Amy, Elisabeth, and Iris

Preface

THIS BOOK, LIKE MOST BOOKS, was conceived in a manner that was both intentional and serendipitous. In the mid 1980s I became intrigued, along with many other historians of science, by the problem of studying scientists' daily work and practices. I was struck by how little had been written on this aspect of science, either by historians or by scientists themselves. No less striking at the time was how scholars everywhere had begun to use the language of craft and practice, although it was not evident that they were using it in the same way or that it was anything more than a gloss on traditional literary modes. Perhaps scholars were simply repositioning themselves in anticipation of a new fashion—a familiar phenomenon in the volatile world of postmodern academic disciplines, especially the social sciences. The question was, Would the intellectual benefits of studying scientific practice match the rhetorical promise? Would the study of material culture and craft practice become a broad and productive new genre or remain an embellishment of familiar intellectual, institutional, and disciplinary histories? Experience would reveal the answers to these questions, and I decided on a case study of biochemical genetics. I had some years of practical laboratory experience in that field myself, and I had already collected material for a study of George Beadle's intellectual history, which I hoped could be diverted to new uses.

My involvement with *Drosophila* was more serendipitous. As a former chemist and molecular biologist I felt more comfortable writing about simple creatures like *Escherichia coli* or *Neurospora* than about those with a more complex biology. I knew that Beadle and Boris Ephrussi had worked with *Drosophila* in T. H. Morgan's Caltech fly group, but I assumed that would be a preliminary to the real story of Beadle and Edward Tatum's subsequent work on *Neurospora*. As will happen with storytelling, however, my characters took charge of the project. I began to get interested in the genetic work that was being done in the fly group in the 1930s, especially Theodosius Dobzhansky's work on the genetics of speciation in natural populations. It seemed quite remarkable that two such novel and productive modes of experimental practice should have been invented in the same place at the same time. I wondered why. I felt out of my depth in *Drosophila* genetics, but was persuaded by my colleague Mark Adams to contribute a study of Dobzhansky's role in the Caltech fly group to an upcoming conference in Leningrad. From there it was but a short step to a full-scale study of how *Drosophila,* that extraordinary little beast, had become such an important presence in the landscape of modern experimental biology.

Some of my original questions were resolved in the course of this experiment. It became clear that the material culture and craft practices of science are not just another specialized subject from which the cream will soon be skimmed. Rather, they are basic behavioral categories, which have the potential to transform (but not replace) traditional accounts of intellectual, institutional, and social history of science. Material culture and craft practices enable historians to use their primary sources, the literatures of science, in a new way. They invite comparative studies across scientific disciplines, offering a generation of historians of science choice subjects for dissertations and books. The study of laboratory and field practices is in a sense the equivalent for historians of science of *Drosophila* and genetic mapping—an endlessly productive machine for generating significant problems.

The most important thing that I learned from this experiment, however, is that writing about the material culture and craft practices of science is fun. Nothing I have written before has gone more smoothly and given me more pleasure than this study of *Drosophila* and the drosophilists. It was a liberating experience, in part perhaps because studying material culture and craft is an especially mobile mode of historical practice. Intellectual history, in contrast, requires a vast investment in arcane knowledge and thus tends to tie historians permanently to particular sciences. The study of institutions similarly tends to tie historians to particular national cultures. In pursuit of the history of tools and practices, historians can forage more widely and enjoy a more varied intellectual diet.

It is also liberating for producers of monographs to know that their products may be accessible to a wider audience of readers and users, including scientists and people involved in science administration. What we historians of science share with our subjects is not, after all, our respective cultural products, which are very different, but the process of intellectual production itself. Historians, like scientists, are inventors of machines that generate problems. Professional workers should like to read and study varieties of professional work, just as middle-class Victorians liked to read and write stories about their own lives and cultural strivings. Storytelling is slowly coming back in fashion among historians of science, and the working lives of scientists make the best stories.

I would like to thank the following friends and colleagues for their assistance. William Provine generously made printed and unpublished material available and gave helpful criticism on evolutionary genetics. Daniel Alexandrov and Mark Adams set me straight on the Russian school of population genetics. I am especially grateful to Guil Winchester for sharing with me rich materials from the Otto Mohr papers, which she discovered. I am once again indebted to the knowledgeable and friendly archivists of the American Philosophical Society, the Lilly Library of Indiana University, the University of Illinois, the Carnegie Institution of Washington, the California Institute of Technology, and the Rockefeller Archive Center. Jill Schultz Frisch and Sophie Dobzhansky Coe generously made family photographs available. The editors of *History of Science, Historical Studies in the Physical and Biological Sciences,* and *Journal of the History of Biology* gave permission to publish material that first appeared in those journals. Judith Goodstein conceived the title in a flash of inspiration during a party given by Dan and Bettyann Kevles. (She could think of a livelier title than *The Drosophilists,* she promised, and proceeded to do so on the spot.) Finally, Frances Kohler (though not without complaint) smoothed and clarified my prose and pruned away crotchets that I had ceased to notice—she made it better.

The Nature of Experimental Life

T HIS BOOK IS ABOUT THE MATERIAL CULTURE and way of life of experimental scientists. It is also about a particular and famous community of experimental biologists, the _Drosophila_ geneticists, and their no less famous co-worker, the fruit fly. Few laboratory creatures have had such a spectacularly successful and productive history as _Drosophila._ It first entered laboratories about 1900, revealed its talent for experimental genetics to Thomas Hunt Morgan and his students at Columbia University in the early 1910s, and after some ups and downs in status is still going strong almost a century later. If not the first "standard" laboratory creature, _Drosophila_ is certainly representative of the type. So, too, is Morgan's original fly group an archetypical experimental community, and the period of its life cycle, from about 1910 to the early 1940s, more or less demarcates the beginning and end points of this book.

A great deal has been written about the "Drosophilists," as they called themselves: in a heroic vein by the last of the founding generation and their students, about twenty years ago, and again in recent years in a more critical and revisionist vein.[1] Despite some ideological

1. Garland E. Allen, "T. H. Morgan and the emergence of a new American biology," _Quar. Rev. Biol._ 44 (1969): 168–88; Allen, "The introduction of _Drosophila_ into the study of heredity and evolution, 1900–1910," _Isis_ 66 (1975): 322–33; Allen, _Thomas Hunt Morgan: The Man and His Science_ (Princeton: Princeton University Press, 1978); Elof A. Carlson, "The _Drosophila_ group: The transition from the Mendelian unit to the individual gene," _J. Hist. Biol._ 7 (1974): 31–48; Carlson, _Genes, Radiation, and Society: The Life and Work of H. J. Muller_ (Ithaca: Cornell University Press, 1981); Daniel J. Kevles, "Genetics in the United States and Great Britain, 1890–1930: A review with speculations," _Isis_ 71 (1980): 441–55; Jan Sapp, "The struggle for authority in the field of heredity, 1900–1932: New perspectives on the rise of genetics," _J. Hist. Biol._ 16 (1983): 311–42; Sapp, _Beyond the Gene: Cytoplasmic Inheritance and the Struggle for Authority in Genetics_ (New York: Oxford University Press, 1987); Jonathan Harwood, "National styles in science: Genetics in Germany and the United States between the world wars," _Isis_ 78 (1987): 390–414; Harwood, _Styles of_

Chapter One

differences, these various historical schools share a common concern
with the conceptual and disciplinary dynamics of genetics—how competing genetic theories and visions of the field were devised, received,
and contested. The richness of this literature makes *Drosophila* genetics
an inviting case on which to try out a different historical approach, one
focused on practices and material culture. This is not to say that the
history of ideas is unimportant, only that the history of the material
and human side of experimental life can be equally productive.

The material culture of experimental scientists has been surprisingly neglected by historians. Although a substantial body of literature
on instruments and laboratories is now accumulating and the subject
seems certain to be popular in the 1990s, the study of the material and
working cultures of experimental workplaces is still very much in its
infancy. Themes and approaches are diverse, and it is not yet evident
which ones will be the most fruitful.[2]

A distinctive feature of this case study is its focus on an experimental organism: *Drosophila*, that crucial bit of material culture upon which
generations of geneticists have come to depend for their careers and
livelihoods. Why do some organisms, like *Drosophila*, become cosmopolitan, "standard" species of laboratory creatures—cornucopias of
productive methods, concepts, and problems—while others do not?

Scientific Thought: The German Genetics Community, 1900–1933 (Chicago: University of Chicago Press, 1993); and Richard M. Burian, Jean Gayon, and Doris
Zallen, "The singular fate of genetics in the history of French biology, 1900–
1940," *J. Hist. Biol.* 21 (1988): 357–402.

2. See especially Peter Galison, "Bubble chambers and the experimental
workplace," in Peter Achinstein and Owen Hannaway, eds., *Observation, Experiment, and Hypothesis in Modern Physical Science* (Cambridge: MIT Press, 1985),
pp. 309–73; Sharon Traweek, *Beamtimes and Lifetimes: The World of High Energy
Physics* (Cambridge: Harvard University Press, 1988); and Hannaway, "Laboratory design and the aim of science: Andreas Libavius versus Tycho Brahe," *Isis*
77 (1986): 585–610. For an up-to-date bibliography see Adele Clarke and Joan
Fujimura, "What tools? Which jobs? Why right?" in Clarke and Fujimura, eds.,
The Right Tools for the Job: At Work in Twentieth-Century Life Sciences (Princeton:
Princeton University Press, 1992), pp. 3–44. Some useful collections of case
studies are David Gooding, Trevor Pinch, and Simon Schaffer, eds., *The Uses of
Experiment* (Cambridge: Cambridge University Press, 1989); Homer Le Grand,
ed., *Experimental Inquiries: Historical, Philosophical, and Social Studies of Experimen-
tation in Science* (Dordrecht: Kluwer Academic, 1990); and Wiebe E. Bijker,
Thomas P. Hughes, and Trevor Pinch, eds., *The Social Construction of Technological Systems* (Cambridge, Mass.: MIT Press, 1987).

Why do some modes of experimental production flourish and spread around the world while others languish in local obscurity? What qualities made the partnership between fly and fly people so effective? I will seek answers to these questions in the process of experimental production, in the biological and technological nature of *Drosophila,* and in the customs and practices of the drosophilists. I hope to persuade readers of this book that experimental sciences have been shaped by their material cultures: by the practical imperatives of choosing organisms, constructing tools, and making experiments work.

In one respect this account is somewhat at odds with what has been the main line of work on experimental practice: namely, in its lack of attention to issues of epistemology and "social construction." It is a historical fact that the first scholars to study experimental practices systematically were not historians but philosophers and sociologists, who sought in scientists' workaday behavior evidence to clinch their demolition of naturalistic epistemologies of scientific knowledge. (Scientists give facts an appearance of natural inevitability by veiling the process of their construction, so sociologists seek to deconstruct facts by unveiling that process and showing how messy and contingent it really is.)[3] However, this strong connection between relativist epistemology and studies of laboratory practice was itself historically contingent upon the peculiar ecology of science studies in the 1980s. There are other, equally good reasons to do empirical studies of experimental practices. For some, myself included, it is the production process itself that is the object of interest—the material culture, social conventions, and moral ordering of experimental production.

There are, in fact, ample precedents in the science studies literature for a realist, production-oriented treatment of experimental practices. For example, in their pioneering ethnography of laboratory life,

3. Bruno Latour and Steve Woolgar, *Laboratory Life: The Social Construction of Scientific Facts* (Beverly Hills: Sage, 1979), esp. ch. 5. Latour and Woolgar pointedly dropped the word *social* from the subtitle in the second edition (Princeton: Princeton University Press, 1986). See also Karen Knorr-Cetina, "Tinkering toward success: Prelude to a theory of scientific practice," *Theory and Society* 8 (1979), 347–76; Knorr-Cetina, R. Krohn, and R. Whitley, eds., *The Social Process of Scientific Investigation* (Dordrecht: Reidel, 1984); Harry M. Collins, *Changing Order,* new ed. (Chicago: University of Chicago Press, 1992); Michael Lynch, *Art and Artifact in Laboratory Science* (London: Routledge & Kegan Paul, 1985); Trevor Pinch, *Confronting Nature* (Dordrecht: Reidel, 1986); Andrew Pickering, *Constructing Quarks: A Sociological History of Particle Physics* (Chicago: University of Chicago Press, 1984); and Pickering, ed., *Science as Practice and Culture* (Chicago: University of Chicago Press, 1992).

Bruno Latour and Steve Woolgar make the fundamental point that scientists work neither out of pure curiosity nor to win rewards of honor and status, but rather to gain the continuing privilege of working under ideal conditions with ample material and social resources. These scholars see experimental work as driven by a cycle of investment, in which experimental work is turned into publications, the symbolic capital of which is reinvested in new machinery of production, which generates more experiments, and so on. It is the work itself that drives the work—a simple but fundamental point. Latour and Woolgar make a second fundamental point when they propose that experimentalists trust and value most highly results that they can put to use productively in their own experimental work. This pragmatic conception of credibility and truth locates the causes of scientists' behavior in the production process rather than in the realm of theoretical beliefs or professional and political ideologies.[4] These perceptions of experimental work seem to me absolutely right and fundamental, and I take them as axiomatic in this study.

Other foundational concepts of a history of scientific production have been developed by Steven Shapin and Simon Schaffer in their studies of Robert Boyle and his circle in seventeenth century England. Shapin, for example, points to the literary, material, and social "technologies" of experimental workplaces, and notes how they enable communities of practitioners to judge the credibility of experimental knowledge by rules that all agree on. The special literary conventions of scientific publication and the social customs of public witnessing and discourse are as essential for experimental production as the material technology of air pumps and other instruments. Shapin and Schaffer make the epistemological point that judgments of true knowledge and personal credibility are social conventions embedded in the material, social, and moral fabric of particular communities.[5] However, their concepts and methods are equally fruitful in understanding the pro-

4. Latour and Woolgar, *Laboratory Life*, esp. chaps. 2, 5. This model is elaborated, though without additional empirical evidence, in Bruno Latour, *Science in Action* (Cambridge: Harvard University Press, 1987). Similar insights into the nature of technical work are vividly displayed in Tracy Kidder's account of a year in the working lives of a group of computer engineers: Tracy Kidder, *The Soul of a New Machine* (Boston: Little, Brown, 1981).

5. Steven Shapin, "Pump and circumstance: Robert Boyle's literary technology," *Soc. Stud. Sci.* 14 (1984): 481–520; Steven Shapin and Simon Schaffer, *Leviathan and the Air Pump: Hobbes, Boyle, and the Experimental Life* (Princeton: Princeton University Press, 1985); and Shapin, "The invisible technician," *Amer. Scientist* 77 (1989): 554–63.

duction process itself: that is, the social and cultural conventions that regulate access to tools and communication networks and distribute tasks and moral authority among the different participants in experimental work. My aim in this book is to identify the material, moral, and social technologies that constituted the drosophilists' work culture. It is to show how a distinctive workplace culture arose out of the process of constructing a "standard" fly, and how this culture shaped the ways in which the fly was used.

Production-oriented concepts have also been used to good effect by sociologists of occupations and work, most notably Adele Clarke, Joan Fujimura, and Susan Star. Their theoretical interest in the constitution of working "social worlds" led them, via a different route from the sociologists of knowledge, to a similar quest for the essential elements of scientific production. Fujimura, for example, has pointed to the importance of doability in experimental practice: all else being equal, experimenters will usually choose problems that are likely to produce significant, usable results. Clarke and others have pointed to the complex organization of modern research and to the importance of managerial and organizational workers in integrating benchwork, fundraising, programmatics, public relations, and so on.[6] These general propositions about the practicalities of doing science have been well grounded in theories of group behavior and have proven their worth in crafting historical case studies. I have appropriated methods and insights freely from these various schools of science studies—though perhaps not always in ways that will please them.

How, then, do we look at the material culture of standard organisms like *Drosophila* and the work cultures of communities like Morgan's fly group? I look at them in three distinct but complementary ways: technologically, biologically, and morally. First, experimental organisms can be understood as technological artifacts that are constructed and embedded in complex material and social systems of pro-

6. Adele Clarke and Elihu Gerson, "Symbolic interactionism in science studies," in Howard S. Becker and Michael McCall, eds., *Symbolic Interactionism and Cultural Studies* (Chicago: University of Chicago Press 1990, 170–214; Joan Fujimura, "Constructing double problems in cancer research: Articulating alignment," *Soc. Stud. Sci.* 17 (1987): 257–93; Fujimura, "The molecular biology bandwagon in cancer research: Where social worlds meet," *Social Problems* 35 (1988): 261–83; Susan L. Star and James Griesemer, "Institutional ecology, 'translations,' and boundary objects: Amateurs and professionals in Berkeley's Museum of Vertebrate Zoology," *Soc. Stud. Sci.* 19 (1989): 387–420; and Star, *Regions of the Mind: Brain Research and the Quest for Scientific Certainty* (Palo Alto: Stanford University Press, 1989).

duction, much as machines are embedded in systems of material mass production. Second, these living instruments also have a biology and a natural history. In the wild they inhabit distinctive ecological niches and have varied relationships with other creatures, including humankind. In the laboratory they enter into yet another kind of symbiotic relationship with humans, as participants in experiments. Experimental workplaces possess distinctive natural histories—like any habitat, wild or domesticated. Finally, experimental groups possess distinctive "moral economies" that regulate authority relations and access to the means of production and rewards for achievement. These moral rules operate within and between groups or practitioners and may perhaps be seen as the operating rules—the cultural software, so to speak—of Latour and Woolgar's reinvestment cycle.

Some readers may find it jarring to think of experimental creatures as technological and biological at the same time. But there is no real incongruity. The workplaces and material cultures of experimental biologists are simultaneously technological, biological, and moral, as are households, farms, forests, or any places where people live and work. These aspects of the work of experimental biologists are not independent but interactive, and they must be brought together to explain how the work of experimental science is done.

Organisms as Technology

What exactly do we mean when we speak of experimental creatures as instruments and technological artifacts? They are not literally machines, of course, but neither is the resemblance merely a matter of analogy or metaphor. Experimental creatures are a special kind of technology in that they are altered environmentally or physically to do things that humans value but that they might not have done in nature. Some are dramatically designed and constructed: the "standard" organisms—*Drosophila,* white mice and rats, maize, *E. coli* or *Neurospora*— which have been reconstructed genetically through generations of selection and inbreeding into creatures whose genetic makeup and behavior are quite different from their natural ancestors'. These are the constructed creatures that most resemble spectrophotometers, bubble chambers, ultracentrifuges, and other physical instruments. The extent of construction varies a good deal, however. With "found" objects like sea urchin eggs, frogs, or primates (including humans), the artifice resides less in physical reconstruction than in the accretion around these creatures of bodies of knowledge about how they behave and how they can be made to do useful tricks in experimental laboratories. The majority of laboratory creatures are, I suspect, technologies in this more limited sense, though we have as yet little basis for generalizing. The transformation of "natural" creatures begins when they enter into their first experiment, and the more productive they become, the more they come to resemble instruments, embodying layers of accumulated craft knowledge and skills, tinkered into new forms to serve the peculiar purposes of experimental life.

What models exist for studying experimental creatures in this way? There are as yet very few studies of experimental creatures as constructed artifacts, apart from this one and Bonnie Clause's history of the standard Wistar rat. Bruno Latour has been urging scholars for some years to regard creatures (and inanimate things) as the equals of human actors, but he does not actually deal with the biology of viruses in his study of Pasteur.[7] Historians of biology may, however, draw on the richer literature on chemical and physical instruments for useful models for similar treatments of the material culture of experimental biology. Case studies from the material culture of physical science and technology demonstrate, for example, how the agendas and social relations of working groups become embodied in the machinery of experimental production.[8] Experimental plants and animals, too, have pro-

7. Bonnie Clause, "The Wistar rat as a right choice: Establishing mammalian standards and the ideal of a standardized mammal," *J. Hist. Biol.* 26 (1993): 329–49; Bruno Latour, "Give me a laboratory and I will raise the world," in Karin Knorr-Cetina and Michael Mulkay, eds., *Science Observed* (London: Sage, 1983), pp. 141–70; and Latour, *The Pasteurization of France* (Cambridge: Harvard University Press, 1988). See more generally Adele E. Clarke, "Research materials and reproductive science in the United States, 1910–1940," in Gerald Geison, ed., *Physiology in the American Context 1850–1940* (Bethesda, Md.: American Physiological Society, 1987), pp. 323–50; and Clarke and Fujimura, eds., *Right Tools for the Job.* For a cultural-studies approach to experimental creatures see Donna Haraway, *Primate Visions: Gender, Race, and Nature in the World of Modern Science* (New York: Routledge, 1989); and Michael Lynch, "Sacrifice and the transformation of the animal body into a scientific object: Laboratory culture and ritual practice in the neurosciences," *Soc. Stud. Sci.* 18 (1988): 265–89.

8. John Lankford, "Amateurs versus professionals: The controversy over telescope size in late Victorian science," *Isis* 72 (1981): 11–28; Joel D. Howell, "Early perceptions of the electrocardiogram: From arrythmia to infarction," *Bull. Hist. Med.* 58 (1984): 83–98; Galison, "Bubble chambers"; Peter Galison, *How Experiments End* (Chicago: University of Chicago Press, 1987); James R. Wright, Jr., "The development of the frozen section technique, the evolution of surgical biopsy, and the origins of surgical pathology," *Bull. Hist. Med.* 59

Chapter One

grams and agendas built into them. Machines have "politics," and so, too, do standard experimental creatures.[9]

Instruments produce nothing by themselves, of course, but only as parts of complex material and social systems of production that enable experimenters to mobilize material resources, socialize recruits, and persuade other workers to accept their results and adopt their methods of production. Historians of technology have been especially alert to the complexity of material production, and historians of biology may find useful models in that work for their own studies of experimental production.[10] A bottle of flies is not of much use for experimental production in itself, but only as part of an assemblage of material instruments, standard recipes and procedures, and working relationships. We need, therefore, to take as our units of study those systems of material, literary, and social technologies in which working groups make substantial investments and which have the property of expanding and diversifying into many lines of work over a period of time. These are the major systems of material culture that constitute the world of experimental science.

(1985): 295–326; David Gooding, "'In nature's school': Faraday as an experimentalist," in Gooding and Frank A. J. L. James, eds., *Faraday Rediscovered* (London: Macmillan, 1985), pp. 105–35; Boelie Elzen, "Two ultracentrifuges: A comparative study of the social construction of artifacts," *Soc. Stud. Sci.* 16 (1986): 621–62; Bruce Hunt, "Experimenting on the ether: Oliver Lodge and the great whirling machine," *Hist. Stud. Phys. Sci.* 16 (1986): 111–34; Timothy Lenoir, "Models and instruments in the development of electrophysiology, 1845–1912," *Hist. Stud. Phys. Sci.* 17 (1987): 1–54; Isobel Falconer, "J. J. Thomson's work on positive rays, 1906–1914," *Hist. Stud. Phys. Sci.* 18 (1988): 267–310; and Robert G. Frank, Jr., "The telltale heart: Physiological instruments, graphic methods, and clinical hopes, 1854–1914," in William Coleman and Frederic L. Holmes, eds., *The Investigative Enterprise: Experimental Physiology in Nineteenth-Century Medicine* (Berkeley: University of California Press, 1988), pp. 211–90.

9. Langdon Winner, "Do artifacts have politics?" *Daedalus* 109 (1980): 121–36; and Susan E. Lederer, "Political animals: The shaping of biomedical research literature in twentieth-century America," *Isis* 83 (1992): 61–79.

10. Thomas P. Hughes, "The evolution of large technological systems," in Bijker et al., *Social Construction of Technological Systems*, pp. 51–82; and Hughes, *Networks of Power: Electrification in Western Society, 1880–1930* (Baltimore: Johns Hopkins University Press, 1983), chap. 1. Latour and Woolgar use the term *culture* in much the way that Hughes and others use *system:* see *Laboratory Life,* pp. 54–55.

Organisms and Natural History

The biology and natural history of experimental creatures is no less important for understanding their modes of life than their history as technology. Experimental plants and animals get into laboratories from nature or, more commonly, from a semidomesticated "second nature" that they already share with humans.[11] Many creatures were already more or less domesticated upon their arrival in laboratories, having served as agricultural producers, civic decoration, or household pets (e.g., peas, primroses, fowl, guinea pigs, dogs). Others, such as mice, rats, weeds, and fruit flies, lived as half-wild commensals in a close but irregular relationship with humankind, as hangers-on in homes, gardens, and city streets—just beyond the thresholds of experimental labs. Still other creatures moved to and fro between domesticated and commensal lives (e.g., pigeons, cats, fish).[12]

It is useful to regard the construction of standard laboratory creatures as a special kind of domestication. The resemblance is obvious between standard laboratory creatures and the highly engineered, standard biological machines of large-scale agricultural technology, such as staple grains and fowl. But an understanding of the biology of domestication also illuminates how creatures get into labs unintentionally, before their potential as technology is recognized. A leading study is David Rindos's application of evolutionary theory to the early, preagricultural stages of domestication. He argues that plants and animals coevolved with hunting-and-gathering peoples, developing the characteristics of morphology and dispersal that were later and deliberately exploited in the domestication process. The establishment of relations between creatures and humankind was not at first intentional, Rindos argues, but the result of automatic biological and evolutionary processes.[13] In the same way, experimental biologists and their creatures

11. For contemporary uses of "second nature" see Donald Worster, "Transformations of the earth: Toward an agroecological perspective in history," *J. Amer. Hist.* 76 (1990): 1087–1106, on p. 1089; and William Cronon, *Nature's Metropolis: Chicago and the Great West* (New York: Norton, 1991), p. xvii. The phrase was apparently coined by Marx.

12. Harriet Ritvo, *The Animal Estate: The English and Other Creatures in the Victorian Age* (Cambridge: Harvard University Press, 1987); and James Serpell, *In the Company of Animals* (Oxford: Basil Blackwell, 1986).

13. David Rindos, *The Origins of Agriculture: An Evolutionary Approach* (New York: Academic Press, 1984). A readable popular treatment of Rindos's ideas

may enter into relationships that cause the behavior of experimenters and the biological nature of plants and animals to change in unintended and unexpected ways—as we will see. In a few cases, as with *Drosophila,* mutual dependence may evolve into the more intense relationship characteristic of standard organisms and experimental mass production.

In this view laboratories and landscapes are not such different places as may appear, though it is conventional wisdom to distinguish sharply between nature and artifice. Laboratories of experimental biology are a distinctive kind of ecosystem, in which creatures live and evolve in symbiotic and commensal relations with humankind. They have natural histories, no less than fields and forests do, and the boundaries between lab and field are active places that may be traversed or occupied in a variety of imaginative ways. Experiments may be carried out in nature but with laboratory instruments and methods (as in, e.g., ecology, astrophysics, anthropology), or in laboratories with material and methods drawn from fieldwork (e.g., population genetics and animal behavior). Simulations of nature may be staged in the laboratory (e.g., cloud chambers, aquaria, insect colonies). Bits of nature may be turned into something like the controlled environment of the laboratory (e.g., quadrats, monkey islands).[14] There is no hard and fast distinction between domesticated and wild places. The boundary zones between lab and field resemble the edge habitats that we humans create so abundantly wherever we go in the natural world.

---

is Stephen Budiansky, *The Covenant of the Wild: Why Animals Chose Domestication* (New York: William Morrow, 1992).

14. Ronald C. Tobey, *Saving the Prairies: The Life Cycle of the Founding School of American Plant Ecology, 1895–1955* (Berkeley: University of California Press, 1981), chap. 3; Joel B. Hagen, "Experimentalists and naturalists in twentieth-century botany: Experimental taxonomy," *J. Hist. Biol.* 17 (1984): 249–70; Christopher Hamlin, "Robert Warington and the moral economy of the aquarium," *J. Hist. Biol* 19 (1986): 131–53; Peter Galison and Alexi Assmus, "Artificial clouds, real particles," in Gooding, Pinch, and Schaffer, eds., *Uses of Experiments,* pp. 225–74; James A. Secord, "Extraordinary experiment: Electricity and the creation of life in Victorian England," ibid., pp. 357–84; Alex Soojung-Kim Pang, "Spheres of Interest: Imperialism, Culture, and Practice in British Social Eclipse Expeditions, 1860–1914," Ph.D. diss., University of Pennsylvania, 1991, chaps. 4, 5; Pang, "The social event of the season: Eclipse expeditions and Victorian culture," *Isis* 83 (1993): 252–77; and Henrika Kuklick, *The Savage Within: The Social History of British Anthropology, 1885–1945* (Cambridge: Cambridge University Press, 1992), pp. 133–49.

Geographers and ecological historians have been epecially vigorous in blurring the distinction between nature and technology. They have shown how many of the landscapes that may seem natural to us were in fact artificially created by humankind with their technologies of fire, plow, and seed.[15] Conversely, they have shown how landscapes that may seem thoroughly tamed and domesticated are nonetheless natural ecosystems. Donald Worster, for example, treats different kinds of agricultural landscape as modes of technological production in his study of the Dust Bowl of the American Southwest. William Cronon similarly writes ecologically about domestic, urban, and technological environments in his remarkable study of Chicago and its connections, via the traffic in agricultural and forest commodities, to the landscapes of the upper Middle West.[16] Landscapes are technologies; technological workplaces have natural histories. Laboratories of experimental biology have that same dual character.

The Moral Economy of Laboratories

Experimental laboratories are places not only of material and social order, but also of moral order, and the moral economy of laboratory life is an equally essential part of experimental production. The term is, of course, borrowed from E. P. Thompson's celebrated 1971 essay on eighteenth-century English bread riots. Thompson meant by *moral economy* the customs, traditions, and moral rules that consumers (especially poor consumers) expected would regulate the market for basic foodstuffs and, in times of dearth, prevent landowners and traders from withholding the essentials of life for the sake of profit. (It was these unstated rules, Thompson argued, that structured behavior in

15. Carl Sauer, "Man's dominance by use of fire," in *Selected Essays 1963–1975* (Berkeley: Turtle Island Foundation, 1981), pp. 129–56; William Cronon, *Changes in the Land: Indians, Colonists, and the Ecology of New England* (New York: Hill & Wang, 1983); and Timothy Silver, *A New Face on the Countryside: Indians, Colonists, and Slaves in South Atlantic Forests, 1500–1800* (Cambridge: Cambridge University Press, 1990).

16. Donald Worster, *Dust Bowl: The Southern Plains in the 1930s* (New York: Oxford University Press, 1979); Worster, "Transformations of the earth"; Worster, "Seeing beyond culture," *J. Amer. Hist.* 76 (1990): 1142–47; William Cronon, "Modes of prophecy and production: Placing nature in history," ibid., pp. 1122–31; Cronon, "A place for stories: Nature, history, and narrative," ibid., 78 (1992): 1347–76; and Cronon, *Nature's Metropolis.*

bread riots).[17] In an extended usage Thompson's concept has proved widely useful to scholars of various kinds. James Scott, most notably, has used the idea to explain the surprisingly peaceable relations between peasant farmers and landowners in South Asia. Although Scott locates moral economy in production and not marketing, as Thompson does, they both deal with nonmonetary obligations and rights of access to the necessities of life, and that makes Scott's a legitimate extension, in Thompson's view.[18]

The concept of moral economy may be similarly extended to the productive life of experimental workplaces. There, too, unstated moral rules define the mutual expectations and obligations of the various participants in the production process: principal scientists and their assistants, mentors and students, well-placed and peripheral producers—researchers who may be collaborators one day and competitors the next. Moral conventions regulate access to tools of the trade and the distribution of credit and rewards for achievement. As the moral economy of eighteenth-century English laborers was rooted in a concrete, historical system of agricultural production and marketing, so are the moral economies of experimental scientists rooted in specific configurations of material, literary, and social technology.

It may seem odd to some, even offensive, to apply to elite, middle-class practitioners a concept that was invented to apply to poor laborers and peasants. Yet if social inequalities are less pronounced and less jarring in experimental laboratories, they do exist, and issues of access and equity are no less real and emotional issues to middle-class professionals than to laborers. The extension therefore seems justified. Indeed, scientific work seems especially well suited to moral analysis, because its reward are not cash (for academic scientists, at least) but authority and access to tools and craft knowledge—the symbolic capital of Latour and Woolgar's credit cycle.

Few historians of science have as yet used the idea of moral econ-

17. E. P. Thompson, "The moral economy of the English crowd in the eighteenth century," *Past and Present* 50 (1971): 76–136, rpt. in Thompson, *Customs in Common* (New York: New Press, 1991), pp. 185–258; and Thompson, "The moral economy reviewed," ibid., pp. 259–315. Thompson contrasts this moral economy with the deregulated, "de-moralized" political economy of free-market capitalism.

18. James C. Scott, *The Moral Economy of the Peasant: Rebellion and Subsistence in Southeast Asia* (New Haven: Yale University Press, 1976); and Thompson, "Moral economy reviewed," pp. 340–49. Thompson warns against stretching moral economy to include all social values or attitudes, lest the concept lose its usefulness as a specific analytical tool: ibid., pp. 338–39.

omy in a systematic way (though the term is beginning to be fashionable). The outstanding exception is Steven Shapin's work on the moral economy of experimental workplaces in seventeenth-century England.[19] Shapin shows how social identities and class relations within these early laboratories were appropriated from the social conventions of an urban gentry, and how these borrowed conventions became permanently embedded in the culture of experimental science. As Thompson locates the moral economy of the crowd in everyday marketing practices, so Shapin finds the moral economy of experimental scientists in the social practices of gentlemen's houses, where the first experimental laboratories were in fact located. Likewise, we will see the moral economy of drosophilists in the places where they work and where their standard laboratory creatures live. We will seek the drosophilists' moral economy in their conventions of recruitment and socialization, collaborative habits of working and publishing, and their distinctive custom of freely exchanging *Drosophila* stocks—the basis of their communal identity and livelihood.

A Social History of Scientists

This book is not a systematic history of *Drosophila* genetics. Indeed, I have done my best to prevent it from growing into one. Rather, it is a thematic study of aspects of drosophilists' working customs that I found relevant to the study of experimental science in general. In selecting topics I have followed the fly, beginning with the story of how *Drosophila* found its way into laboratories and how this wild, highly variable creature was constructed into a standard instrument that could be used for precise, quantitative genetic mapping (chapters 2 and 3). I then turn to the construction around the Morgan group's mapping project of a distinctive human community, the fly group, and of an extended network of fly people who were connected by means of a system of exchanging standard stocks and craft knowledge (chapters 4 and 5).

The emphasis in these first chapters is on the biological and ecological nature of *Drosophila* and on the construction around the stan-

19. Steven Shapin, "The house of experiment in seventeenth-century England," *Isis*, 79 (1988): 373–404; Shapin, "'A scholar and a gentleman': The problematic identity of the scientific practitioner in early modern England," *Hist. Sci.* 29 (1991): 279–327; Shapin, "Who was Robert Hooke?" in Michael Hunter and Simon Schaffer, eds., *Robert Hooke: New Studies* (Woodbridge, U.K.: Boydell Press, 1989), pp. 253–86; and Adi Ophir and Steven Shapin, "The place of knowledge: A methodological survey," *Sci. in Context* 4 (1991): 3–21.

dard fly of a material culture of experiment. A second focus is the creation of the drosophilists' distinctive way of life and the rules of etiquette that tempered generational tensions within the Morgan group and regulated the communal use of standard tools. It is in the exchange network, especially, that the drosophilists' distinctive customs and moral economy are most clearly expressed.

In part 2 I turn again to the standard fly, to examine how the capacities for genetic mapping that were built into it in the 1910s hampered efforts in the 1920s to use it to explore the genetics of development and evolution (chapter 6). Finally, I relate how in the 1930s new experimental systems were invented to reunite classical genetics with development and evolution, by applying to *Drosophila* methods from other disciplines, such as embryology, biochemistry, and natural history (chapters 7 and 8).

These topics do not exhaust the possibilities of a history of experimental organisms and practices, but they do apply generally to other creatures and other disciplines. All laboratory creatures must somehow get across the threshold between nature and lab, and many are constructed once they get there. All standard organisms are associated with complex systems of centers and networks for dispersing standard tools. That is what *standard* means: the things that everyone uses. And the specialized purposes built into standard organisms almost inevitably become confining as research fashions change. Thus, *Drosophila* and the drosophilists are a special case but also, I hope, an exemplary one for historians of biology.

There are real advantages in taking as a unit of historical study those who share a particular organism, rather than those who share a theory, problem, or discipline. I hope that these advantages will become evident to readers. But there are disadvantages, too, which should be pointed out. Following the fly makes it more difficult, for example, to study patterns of competition and emulation among geneticists who work on different organisms—a most interesting problem, which I had reluctantly to set aside for another day. Following the fly also led me away from analyzing the diffusion of *Drosophila* production to secondary and tertiary institutions, since the limiting factors for building programs were not access to standard tools but patronage and local institutional politics. For the same reason I set aside a planned chapter on the experiences of peripheral and underprivileged drosophilists, when it became clear that their disabilities were institutional and not specific to their material culture or moral economy. Those who wish to do intellectual or institutional history may not want to follow flies, mice, or maize, but rather theories, problems, or money.

One final caveat: readers familiar with the history of genetics will note the absence here of any systematic discussion of the major concepts and discoveries of classical *Drosophila* genetics. I go into considerable technical detail about the production process—instruments, procedures, strategies—but not about the products of research. This will disappoint some readers; however, there are excellent books on the intellectual history of genetics, to which they may turn. My aim here is to reveal the nature of experimental work and life to readers who may not be passionately interested in the history of genetics as such.

Indeed, one potential benefit of studying the material culture of scientific work is that this aspect of science may be more accessible to nonspecialists than scientific knowledge. Historians can explain in detail how experimental production works without getting entangled in technicalities. That is because the processes of experimental production are fewer and simpler than their products. Histories of experimental practices and lifestyles may enable historians of science to win a wider audience for their work outside their own speciality without letting go of their special subjects, namely, scientists and their work. The natural history, material culture, and moral economy of science offer the prospect of a general history of science that is not Balkanized by disciplinary specialization.

I am not the only one to entertain such hopes. David Miller recently observed that experimental practices are an ideal subject matter for a newly capacious mode of institutional history that can deal equally well with social organization and intellectual creativity:

> It is not stretching usage too far to describe "experiment" as a scientific institution. The creation and development of the experimental form of life is one of the most central institutional developments in modern science. Recent work, with its focus on the generation of the material, literary and social technologies involved in the experimental form of life, actually provides just that integration of the so-called "life of the mind" and institutional and organizational history which . . . [many historians of science] so much want to see.[20]

This study of *Drosophila* and the drosophilists should be read as an experiment in the kind of history for which Miller calls: a material, cultural, and social history of scientists at work.

20. David P. Miller, "Values redivivus?" *Soc. Stud. Sci.* 22 (1992): 419–27, on p. 424.

PART I

Constructing the System

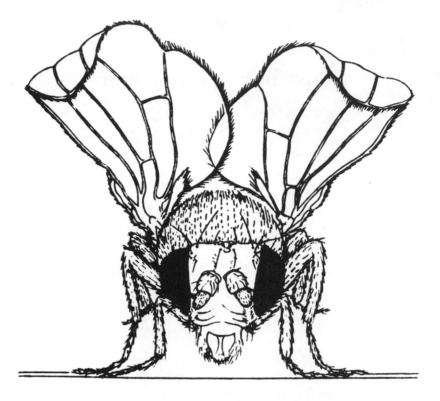

Figure 2.1   *Drosophila melanogaster* (bent-wing mutant). From Morgan,
Bridges, and Sturtevant, "Genetics of Drosophila," p. 215.

TWO

Crossing the Threshold

MEET *DROSOPHILA*, THE FRIENDLY FRUIT FLY. Who has not seen fruit flies congregating about decaying vegetables and fruits in gardens or around garbage bins in public parks? Who has not kept bananas or cantaloupes too long in hot weather and wondered at the sudden appearance in the kitchen of a swarm of the little flies, seemingly by spontaneous generation? Who has not had drosophilas share a glass of beer or cider, or had a microscopic view of them in a high-school or college biology class—and perhaps even bred varieties with oddly colored eyes (not the normal handsome red), extra parts, or misshapen wings? No wild creature, except perhaps the cockroach, is a more familiar companion of humankind. Few inhabitants of experimental laboratories are at once so inconspicuous and so famous. *Drosophila* has been a standard laboratory animal for so long that we no longer think much about how, as a wild creature, it was brought into an alien environment and there transformed into an instrument of scientific production. How did it join unequivocally the multitude of other pets, companions, slaves, and hangers-on of humankind? How did it become a part of that special "second nature" that biologists inhabit?

When fruit flies crossed the threshold of the experimental laboratory, they crossed from one ecosystem to another quite different one, with different rules of selection and survival. Once in the lab they were physically reconstructed and adapted to experimental uses. They were, however, active players in the relationship with experimental biologists, capable of unexpectedly changing the rules of experimental practice. Fruit flies were not just molded like putty by geneticists and put through their paces. Even as the standardized instrument of genetics they had the capacity to change and frustrate drosophilists' plans and change the purposes for which they had originally been brought into the lab. They drove out other favorite creatures of experimental biologists. We need to see the relation between *Drosophila* and drosophilists as an interactive and evolving symbiosis within the special ecological spaces of experimental laboratories.

The natural history of the fly is not quite as rational as it is sometimes portrayed. *Drosophila* was not first brought into the laboratory to do genetics, as the usual story would have it. Rather, I argue, it was brought indoors because its habits and seasonal cycle were well-suited to the needs and seasons of academic life, the need, for example, to give students hands-on experience in experimental manipulations in a limited period of time and with minimal expense. Once in the lab, however, *Drosophila* found a new and more important use in studies of experimental evolution, and in the course of that line of work the little fly revealed an unexpected and very remarkable capacity for experimental heredity and genetics, which soon made it and its human symbionts famous. One mode of experiment led to another in a fast-paced succession that quite unexpectedly transformed the practices in Morgan's laboratory and, eventually, the practices of experimental biologists everywhere.

What, then, was *Drosophila* before it entered the laboratory? How did it cross that ecological threshold, and how, having gained a niche in a new ecosystem, did it establish itself as a dominant species, upon which the fly people came to depend absolutely for their professional livelihoods?

The Natural History of *Drosophila*

The genus *Drosophila* of the family Drosophilidae is made up of over nine hundred species; these inhabit every corner of the world except the driest deserts and the coldest tundras. Drosophilas need moisture (the name means "dew lover"), a temperate climate, and yeasts, which is what they feed on. ("Fruit fly" is thus a misnomer; "yeast fly" would be more apt.) Where there is decaying vegetation or fruit to support colonies of yeasts, there are fruit flies. Some species are highly local and specialized, feeding exclusively and shyly around their favored forest fungi, flower pollen, water plants, fruits, or tree sap. Such specialists tend to be rare and seldom seen except by field naturalists. Other species are less finicky eaters and forage widely and opportunistically. They are thus more ubiquitous and visible. Specialization is the rule: about 90 percent of all *Drosophila* species are endemic to only one of the six great ecological regions of the world. Only eight species groups inhabit all six regions.[1] These cosmopolitan groups were the ones that found

1. John T. Patterson and Wilson S. Stone, *Evolution in the Genus Drosophila* (New York: Macmillan, 1952), pp. 6–7, 48; and Alfred H. Sturtevant, *The Classi-*

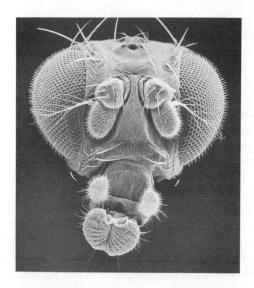

Figure 2.2 *Drosophila melano-gaster* as seen by the scanning electron microscope. Courtesy of F. Rudolf Turner and Thomas Kaufman, Indiana University Department of Biology, Bloomington.

their way into biologists' laboratories. Experimental biologists had no reason to venture from their laboratories into forests to seek out rare species and bring them back alive. The cosmopolitan species were in a position to make the move; they were all about just across the threshold of the lab. Leave a ripe banana on a window sill for awhile, then put it in a jar and wait for the drosophilas to hatch. Chances are you will have *D. melanogaster*, the standard laboratory fly. You can tell it by the black bottom of the males (hence its name).

The cosmopolitan species of *Drosophila* are seasoned and hardy travelers. Blown about by wind and storm, they have been collected from airplanes 3,000 feet in the air. But mainly they have traveled with humankind, of which they are devoted hangers-on, especially *D. melanogaster*. The distribution of this species group suggests that it evolved in Southeast Asia and then spread around the world. It probably got to the New World from the Mediterranean and North Africa via slave ships, and from the Caribbean it traveled north along the trade routes of the international traffic in rum and sugar and of the new traffic in bananas and fresh fruits that burgeoned after the American Civil War. It is no accident that *D. melanogaster*, or *D. ampelophila* as it was first called, was first sighted in the eastern cities of North America in the

fication of the Genus Drosophila, with Description of Nine New Species, University of Texas publication no. 4213 (Austin, 1942), pp. 5–51.

1870s.[2] *Melanogaster* is the great hitchhiker among the drosophilas, as *Homo sapiens* is the great traveler among the primates; wherever its human symbionts have migrated, *D. melanogaster* has gone along for the ride.

It was agriculture, most notably fruit growing, that made possible this symbiosis between *Drosophila* and humankind—agriculture and its associated preserving and fermenting technologies. This connection is clearly revealed in *Drosophila's* common names; vinegar fly, wine fly, pickled-fruit fly, pomace fly (pomace is the apple mash left over from cider pressing). *Ampelophila* means lover of grape vines. Drosophilas hang around humans because their hosts are such prodigious consumers of fermented fruits and such prodigious producers of organic waste and garbage. Drosophilas flourish in vineyards and orchards, breweries, cider mills, pickle works. They breed prodigiously in the piles of discarded and rotting fruits and vegetables that surround citrus orchards or fruit wharves. In the cold of North American cities and towns *Drosophila* can winter over in root cellars, fruit stores, and restaurants. Most cosmopolitan species are fruit feeders and use the north-south trade in tropical fruits to reoccupy the North each year, congregating wherever fruits accumulate at centers of production, distribution, processing, and disposal. The hardiest species of these hitchhikers thrive around garbage pails in parks and on municipal dumps.[3] These cosmopolitan species of *Drosophila* were founding partners in the complex ecological system of human agriculture, along with domesticated plants and animals, pests, and parasites.

Drosophilas were not domesticated creatures, obviously, but they were specialists that had adapted to the domestic economy of humankind. It is no accident that experimental biologists eventually took up with *Drosophila*, nor that the cosmopolitan *D. melanogaster* became the standard laboratory fly. Of all the species of *Drosophila, melanogaster* was the most likely to be associated with urban academic biologists, the most at home among concentrated humanity. It was the most accustomed to crossing the threshold between indoors and out, the most opportunistic and versatile at foraging, and thus the most likely to find its way into a banana-baited trap and an experiment.

2. Patterson and Stone, *Evolution*, pp. 52–55; and T. H. Morgan, C. B. Bridges, and A. H. Sturtevant, "Genetics of *Drosophila*," *Bibliographia Genetica* 2 (1925): 5–6.

3. Patterson and Stone, *Evolution*, pp. 91–92; and Warren P. Spencer, "Collection and laboratory culture," in Milislav Demerec, ed., *Biology of Drosophila* (New York: Wiley, 1950), pp. 547–49.

Experimental Biologists and New Organisms

For *Drosophila* to become a domesticated creature it had first to cross the threshold of the laboratory—to venture from its natural home in the gardens and dustbins of humankind into the domain of the creators of a second, experimental nature. The question is, When and why was Drosophila brought indoors, not accidentally, as a pest to brushed aside or squashed, but as a fellow laborer, to be disciplined and worked?

As Garland Allen has shown, T. H. Morgan was one of half a dozen or so people who began to employ *Drosophila* for experimental work in the early 1900s. The first to use the fly were William E. Castle and his students at Harvard University, in 1901, and most other early drosophilists got the idea directly or indirectly from them. William J. Moenkhaus brought *Drosophila* into his laboratory at the Indiana University Medical School in 1903, having got the idea from Castle. Around 1905 Moenkhaus persuaded the entomologist Frank E. Lutz that it might be a good organism for his work on experimental evolution at the Carnegie Institution's Station for Experimental Evolution at Cold Spring Harbor, Long Island. It was probably Lutz who introduced Morgan to *Drosophila* about 1906, and in 1907 Morgan passed it along to Fernandus Payne, who took it back with him to Indiana University. Morgan's former student Nettie Stevens carried out cytological work on the fly at Bryn Mawr at about the same time, and a few years later at the Rockefeller Institute Jacques Loeb tried to get mutations in *Drosophila,* apparently independently of Morgan's work.[4]

The distribution of these first drosophilists is not random but marks the nodes of an expanding network of American experimental zoologists. Both Moenkhaus and Lutz were students at the University of Chicago when Loeb was there in 1900–1903. Charles Davenport, in whose group Lutz worked, was Castle's colleague at Harvard's Museum of Comparative Zoology, a hothouse for experimental zoologists in the late 1890s. Davenport brought Lutz to Cold Spring Harbor in 1904 as part of his plan to make the Carnegie laboratory into a center of

---

4. Garland E. Allen, "The introduction of *Drosophila* into the study of heredity and evolution, 1900–1910," *Isis* 66 (1975): 322–33; A. F. Blakeslee to T. H. Morgan, 22 May 1935, THM-CIT 1.8; J. W. Wilson to A. H. Sturtevant, 20, 28 Sept. 1965, and F. Payne to Sturtevant, 16 Oct. 1947; all in AHS 3.19; Sturtevant to Payne, 29 Sept. 1947, FP; W. E. Castle, "Memorandum on the beginning of genetic studies at Harvard University (1901–1915)," n.d., UCDG; and H. J. Muller to N. W. Pirie, 3 Jan. 1964, AHS 3.19. Castle was put on to *Drosophila* by his lab mate, the entomologist C. W. Woodworth.

experimental heredity. Moenkhaus studied at both Harvard (1896–98) and Cold Spring Harbor (1904 or 1905). The Indiana University zoologists fed recruits to E. B. Wilson and Morgan at Columbia and served as a secondary center of the experimental movement in the Middle West.

Experimental zoologists' interest in *Drosophila* was symptomatic of a general movement to go afield for new experimental materials. From the late 1890s to the early 1910s many new organisms were brought into laboratories and tried out as experimental creatures. Experimental biology grew explosively in these decades, especially in the United States, and was rapidly dividing into different fields or disciplines— general physiology, animal behavior, heredity, experimental evolution, ecology, and so on. These experimental subjects displaced natural history in colleges and universities, and as hands-on laboratory teaching became a standard feature of undergraduate training, a new generation of biologists bypassed natural history to participate in the new fashion for experiment. In 1906 only 8 percent of the papers in *American Naturalist* were experimental; six years later the figure was 53 percent.[5]

These dramatic changes in the way biology was practiced and taught were powerful incentives to explore the potential of new organisms for experimental production. Specialization encouraged biologists to search for creatures that were especially suited to their particular varieties of experiment. In the absence of established mainstreams and dominant organisms, eclectic foraging for novel organisms was an excellent strategy for making careers. Domesticating unfamiliar creatures to life in the laboratory required arcane knowledge and novel craft skills; it was thus quite an effective way for experimentalists to stake out and protect (temporarily) special lines of work. Indeed, it may be generally true that biologists work on diverse organisms in part to contain or avoid intraspecific competition, just as animals in nature

5. Garland E. Allen, *Thomas Hunt Morgan: The Man and His Science* (Princeton: Princeton University Press, 1978), pp. 32–35; Allen, "Naturalists and experimentalists: The genotype and the phenotype," *Stud. Hist. Biol.* 3 (1970): 197–209; Jane Maienschein, *Transforming Traditions in American Biology, 1880–1915* (Baltimore: Johns Hopkins University Press, 1991); Keith R. Benson, Jane Maienschein, and Ronald Rainger, eds., *The Expansion of American Biology* (New Brunswick, N.J.: Rutgers University Press, 1991); Rainger, Benson, and Maienschein, eds., *The American Development of Biology* (Philadelphia: University of Pennsylvania Press, 1988); and Philip Pauly, "The appearance of academic biology in late nineteenth-century America," *J. Hist. Biol.* 17 (1984): 369–97.

avoid competition by developing specialized ranges and feeding habits. The natural history of creatures in experimental laboratories favors dispersal and diversity, as does the natural history of creatures in the wild. For experimental biologists, symbiosis with different creatures provides elbowroom for individual work within a framework of shared values and problems. An earlier generation of experimental morphologists had resorted to marine biological stations, building careers by exploiting the novel and bizarre richness of marine fauna. In the early 1900s a new generation of experimentalists likewise ransacked wild and civilized nature for creatures that could be transformed into instruments of experimental work.

Although *Drosophila* is famous as a genetic instrument, its early uses were quite varied. At Harvard especially, general physiologists and students of animal behavior took advantage of the genus's marked photo- and chemotropisms (fruit flies move toward light when disturbed and follow vapor trails of alcohol to fermenting fruit). Experiments were also done to detect effects of radium emanations on rate of growth.[6] But most of the early work on *Drosophila* dealt with heredity and evolution.

Few species of biologist foraged more widely for new organisms than those that worked on experimental heredity and evolution, and it is easy to see why. The innumerable varieties of domesticated plants and animals constituted a rich, "natural" resource for experiments on inheritance. More precisely, they were a partly domesticated resource, created over a span of centuries and millenia by breeders and fanciers who selected and preserved rare sports and mutants. Darwin cultivated breeders and fanciers to get evidence to support his theory of natural selection; experimentalists consulted them less, with the notable exception of Gregor Mendel.[7] A vast accumulation of experimental material thus went unused by experimentalists until the rediscovery of Mendel's work in 1900, when it became available virtually overnight. Like placer gold or petroleum deposits, old-growth white pine forests or virgin

6. F. W. Carpenter, "Reactions of the pomace fly," *Amer. Nat.* 39 (1905): 157–71; W. M. Barrows, "Reactions of the pomace fly to odorous substances," *J. Exp. Zool.* 4 (1907): 515–37; and E. D. Congdon, "Effects of radium on living substance, I: The influence of radiations of radium upon the embryonic growth of the pomace-fly *Drosophila ampelophila,* and upon the regeneration of the hydroid *Tubularia crocea,*" *Bull. Museum Compar. Zool.* 53 (1912): 345–58.

7. James A. Secord, "Nature's fancy: Charles Darwin and the breeding of pigeons," *Isis* 72 (1981): 163–86; and Secord, "Darwin and the breeders: A social history," in David Krohn, ed., *The Darwinian Heritage* (Princeton: Princeton University Press, 1986), pp. 519–42.

loess prairies, this slowly concentrated resource was suddenly opened for exploitation in a new kind of production system. So, too, were the accumulated genetic traits of domestic plants and animals a windfall for experimental biologists. Selected not for their ability to survive in the wild but for domestic uses in agriculture, sport, or family life, this accumulated resource was put to new uses in the production of experimental knowledge. For the first generation of neo-Mendelians, foraging quickly and staking an early claim to a particular organism was a strategy of choice, offering the most likely prospect of payoff from a modest investment of labor. It was the equivalent of a gold rush or oil boom.

No one displayed this eclectic foraging habit more dramatically than T. H. Morgan. Restless and energetic, Morgan never specialized for very long and was happiest when he had several lines of work and different organisms in play. He loved to explore a new system, work it for a while, and then move on. No one among his contemporaries carried on more varied researches. In the early 1900s he had five major lines of work going, each with its particular experimental animal. Besides his morphological work with such marine invertebrates as *Tubularia*, he worked on regeneration in amphibia and fish. For studies of development he used frog or toad eggs, and for sex determination, *Phylloxera*, a relative of aphids with a usefully odd sex life. For Mendelizing, the different varieties of domesticated and wild mice and rats seemed the best bet, as insects like *Drosophila* did for studies of mutation and evolution. And those were just the major lines. Morgan also centrifuged mollusk eggs, hoping to get aberrant distributions of chromosomes, and tried (in vain) to find a reliable source of Western song sparrows, which he hoped to resettle in the Bronx Zoo for a study of adaptive variation. When a graduate student was told by his doctor to find outdoor work, Morgan got the idea of sending him to Hawaii to collect the multitudinous species of land snail that had proliferated there in isolated coastal valleys, for a study of variation and speciation in the wild. He kept a small colony of pigeons for breeding work (and later guinea pigs and fowl, too) and avidly attended fancy dog and poultry shows. Sturtevant reckoned that Morgan worked with over fifty different species.[8] "That man has more irons in the fire," Frank Lutz quipped, "than an ordinary man has coals." It was true; Morgan himself

8. A. H. Sturtevant, "Thomas Hunt Morgan," *Biog. Mem. Nat. Acad. Sci.* 33 (1959): 283–325, on pp. 296–98; and Morgan to C. B. Davenport, 30 Apr. 1904, 22 July 1905, 25 June, 4 July 1906 (sparrows); 23 Jan., [ca. 27] Jan. 1905 (snails); 5 June 1905, 5 Feb. 1907 (shows); all in CBD.

poked fun at his habit. "I am doing many things as usual," he told his friend Hans Driesch, "also as badly."[9] But as a foraging strategy it paid off.

Other experimentalists also foraged widely but usually within a single kingdom of the natural or seminatural worlds. In 1910 Castle maintained colonies of 400 rabbits, 700 guinea pigs, 500 mice, 1,000 rats, 400 pigeons, and 8 dogs. He had once had yellow cats, but they had "up and died," and the frogs did not count. Frank Lutz was similarly far ranging within the world of insects, working on *Gryllus* (crickets) and, as time allowed, *Crioceris, Hyphantria,* and *Spilosoma.*[10] Morgan's eclectic interest in experimental creatures was somewhat extreme, but it exemplified a generational style.

Morgan's habit of trying out new organisms was a conscious strategy for keeping up with or insinuating himself into fast-moving research fronts. He was a great reader or forager of the literature but felt unable to judge his colleague's experiments unless he knew from firsthand experience how the experiments had been done. In the case of sex determination, for example, he made himself do "this painful cytological work in order to know at first hand what the cytological evidence means."[11] He took up breeding experiments with mice and rats in 1905 in order to assess the worth of Lucien Cuenot's published work. He suspected that the neo-Mendelians' baroque and ever-changing genetic formulas were artificial concoctions, but he felt unable to criticize until he had concocted some himself. At first he had to seek reassurance from Davenport that he was not just cluttering the literature with more Mendelian nonsense.[12] But confidence came with mastery of the craft skills of mouse breeding, and Morgan was soon gathering material from the field with his usual enthusiasm. He enlisted his friend Charles Zeleny to send wild mice from the environs of Urbana, Illinois, and got William Tower and other friends to send mice from Mexico and California. Morgan himself collected in Florida when

9. F. Lutz to C. Zeleny, 30 Oct. 1908, CZ 2; and Morgan to H. Driesch, 23 Oct. 1905 (see also 15 Sept. 1907), THM-APS. See also Morgan to E. G. Conklin, 2 July 1904, EGC.

10. W. E. Castle, report, 1 Sept. 1910, CBD (also in CIW); and Frank B. Lutz, "Breeding strains of insects," *CIW Yearbook* 6 (1907): 78.

11. Morgan to Driesch, 30 Jan. 1909, THM-APS.

12. Morgan to Davenport, 5 June 1905, 5, 9, 12, 14, 24 Dec. 1905, 24 Jan., 5 Apr., 20 Aug. 1906, 15 Apr. 1907, CBD; Morgan to Driesch, 23 Feb., 16 Aug., 15 Sept. 1906, 27 Nov. 1907, 20 Jan. 1908, THM-APS; and Allen, *Morgan,* pp. 132–36.

local collectors did not come through. Domesticated mice could usually be bought from fanciers (though the source of a desirable yellow
strain of mice proved hard to track down) and were easier to breed
than wild varieties.[13] Thus, what began as an exercise in self-education
became a regular part of Morgan's research program: regeneration,
sex determination, *Entwicklungsmechanik,* and heredity—each with its
particular experimental organism.

In this way the domestic ecosystem of the lab became more diverse
in its inhabitants and more finely subdivided into regions of specialized
experimental practice. The boundary between field and lab became
more permeable, and traffic across it picked up. *Drosophila* was just one
of many creatures that hitchhiked in.

Experimental Evolution: *Drosophila* and Its Competitors

Drosophila's crossing of the threshold between field and lab was unre-
markable, it seems. Morgan did not write about the fly in his letters to
colleagues, as he did about his mice and other new creatures. Neither
Castle nor Lutz thought to mention it in their annual reports to the
Carnegie Institution. Perhaps *Drosophila* was too available and cheap,
too humble a creature to fuss over, not requiring the attention due
expensive, already-domesticated creatures like guinea pigs. So why was
Drosophila domesticated?

It is usual to answer this question by pointing to Drosophila's now-
familiar advantages for large-scale breeding work. These little creatures
breed fast, completing a reproductive cycle within ten days to three
weeks, depending on the species. And they have large families: A single
female can produce from one to four hundred offspring routinely and
up to a thousand if pushed. These demographic qualities give *Drosophila* an obvious advantage for experimental breeding. Plants, in contrast,
have large families but long life-cycles, typically one per year. Rodents
produce twenty generations in a year but have small families. Also, drosophilas have an unusually small complement of chromosomes (four
to seven), though that was not recognized as an advantage at first.[14]

But *Drosophila* also had disadvantages for experimental work on heredity and evolution. Since it was not a domesticated animal, it had no
conveniently distinct and visible characters already built into it. Nature

13. Morgan to Zeleny, 30 Sept., 11, 12 Nov. 1906, 2 May 1907, CZ; and
Morgan to Davenport, 4 July 1906, CBD.

14. Metz, report to Davenport, 25 Sept. 1914, CBD; and Sturtevant, "On
the choice of material for genetic studies," 10 May 1969, pp. 4–6, AHS 19.

provided no albino, yellow, or black drosophilas to compete with mice in the ecosystem of the lab, no wrinkled and round drosophilas to compete with peas. No natural mutants had been collected and lovingly preserved by fanciers. Most natural variations in *Drosophila* consisted of minute differences that blended together along a continuous spectrum and varied wildly with environmental conditions. In the early 1900s the creatures best suited to neo-Mendelian work were the semidomesticated creatures of the fanciers and breeders—the inhabitants of second nature. Given the abundance of such species, there was little reason to choose wild drosophilas for studies of heredity. They had no pedigrees and could not compete with organisms that happened to be preadapted to life indoors in laboratories.

Most important, *Drosophila* had no prestige, no standing among the families of wild or domestic creatures. They were seen as pests. Living in disagreeable and unhealthy places, in dumps, garbage piles, and rotting vegetation, they kept bad company and were assumed to have bad habits. Charles Metz warned darkly of species that had a special affinity for toilets and dinner tables.[15] Drosophilas are not good to eat (though we probably eat more of them than we know in overripe fruits), and most people think they do not make very good pets. They were not fancied and bred into striking forms for collectors, or kept in cages on the desks of scholars or groomed for careers as fighters, as crickets were in China. They were not racers and had no homing instinct. Among the most common and insignificant of God's creatures, drosophilas lent no prestige to those who bred them. Even rats and mice had acquired, through domestication and specialized training, a certain standing in middle-class human society. The expense and trouble of maintaining large rodent colonies for research brought notice and respect. Not so the pesky little flies. Appropriately, they entered the laboratory through the back door, as a kind of poor relation to established domesticated creatures.

Neo-Mendelian geneticists did not domesticate *Drosophila* to remedy the deficiencies of mice or primroses, as is sometimes assumed. *Drosophila* turned out to be good for that purpose, but it first entered the laboratory to take part in experiments on evolution and on variation in physiological characters that were crucial to survival in the wild, such as vitality, fecundity, sex ratio, and growth rates. It was an important issue for evolutionary biologists at the time whether such characters were inherited or determined by environment. Thus most of the

15. Metz, "Biologists and the war," n.d. [late 1917], CBD. *D. funebris* does in fact seek food on excrement, but there is little evidence that it spreads disease.

early breeding experiments with *Drosophila* were not neo-Mendelian crosses but biometric studies of complex characters of the whole organism. In such experiments the advantages of speedy and prolific breeding outweighed the absence of domesticated, Mendelizing characters. Even in these experiments, however, *Drosophila* was not at first seen as an organism of choice. Indeed, it was more often a last resort when other creatures failed or were too costly.

Castle, for example, used *Drosophila* only to show that inbreeding had no effect on vigor; it was a negative control for his experiments on the genetics of coat color in mice and guinea pigs. Since these experiments involved inbreeding, it was important to head off criticism that the genetic results were simply a physiological artifact of the method.[16] *Drosophila* was distinctly a makeshift. Castle clearly expected a negative result, and it would not have made sense to do a routine control with his expensive, slower-breeding mammals.

Similarly, William Moenkhaus would have preferred to use higher animals for his experiments on the effect of inbreeding on fertility, sterility, and (later) sex ratio. Fish were his favorite laboratory creatures, and he would have used them except for the fact that he lacked facilities for large-scale breeding. So he tried mice instead, which proved "miserable failures" in his hands, as did willow beetles, which Moenkhaus soon learned produced only one generation per year in Northern latitudes. Only then did Moenkhaus turn to *Drosophila*, as a last resort.[17] Fernandus Payne was also less than enthusiastic about Morgan's suggestion that he raise generations of *Drosophila* in the dark to see if disuse would cause their eyes to degenerate—a project clearly in the experimental evolution mode. (Morgan's suggestion was prompted by the fact that Payne had previously done some work on blind cave fishes.) Payne thought the experiment was unlikely to yield positive results and so decided to work instead with E. B. Wilson, who used insects and other creatures that he knew were apt for cytological work.[18]

16. W. E. Castle, F. W. Carpenter, A. H. Clark, and S. O. Mast, and W. M. Barrows, "The effects of inbreeding, cross-breeding, and selection upon the fertility and variability of *Drosophila*," *Proc. Amer. Acad.* 41 (1911): 729–86; and Castle, "Memorandum on the beginning of genetic studies at Harvard University 1901–1915," n.d., UCDG.

17. W. Moenkhaus, "The effects of inbreeding and selection on fertility, vigor and sex ratio of *Drosophila ampelophila*," *J. Morphol.* 22 (1911): 123–54.

18. F. Payne, "Forty-nine generations in the dark," *Biol. Bull.* 18 (1910): 188–90; and Payne to Zeleny, 7 Oct. 1907, CZ 2.

Frank Lutz likewise took up *Drosophila* as a second string to his main line of work on variation of short-winged and long-winged forms of wild crickets—a good trait for experimental heredity. With *Drosophila* Lutz chose a more physiological problem, namely, whether the duration of the egg, larval, and pupal stages of insect development were specific, inherited characters or whether they were determined by environmental conditions.[19] *Drosophila* did not prove an ideal subject, however. Its larval and pupal growth rates were just too variable to measure quantitatively. "Think of studying the inheritance of a character that can be doubled by a difference of 5 [degrees] temperature," he complained. "A whole winter [has] gone and only an unsatisfactory paragraph in CIW Yearbook to show for it."[20] Lutz was an ambitious and hard-driving man. As Mabel Smallwood, the librarian at the Carnegie station at Cold Spring Harbor, observed: "Occasionally he was as nice as could be but much of the time he was very proud and haughty and would not play with common people at all. He is working hard and is not at all cheerful much of the time."[21] Lutz was not one to tolerate an organism that did not pay off.

Lutz was eloquent about the limitations of his drosophilas. "Bugs, bugs, bugs!" he exploded to his friend Charles Zeleny in 1905. "Gee! but I have *got* them. I daren't even drink soda for fear of the d.t.'s. I see the darn things crawling in my dreams as it is." And they did not improve on further acquaintance: a year later he complained of feeling "covered ten miles deep with vinegar flies (pesky little things)." After another year of fruitless work he wondered why had he let himself be talked into working with them: "Tell Moenkhaus the flies are still flying," he wrote Zeleny, "and cuss him (for me) for getting me started with the pesky little beasts."[22] Doubtless Lutz was just letting off steam, but his expostulations testify to early drosophilists' generally ambivalent feelings about their bottled-up swarms of restless, prolific little flies. (Zeleny, too, sometimes yearned to "leave the pesky Drosophila" for more traditional problems and experimental creatures.)[23]

A few months later, however, the drosophilas in Lutz's cultures unexpectedly "started to do tricks with their [wing] veins"; that is, wing

19. Lutz, "Breeding strains of insects."

20. Lutz to Zeleny, 24 Feb. 1907, CZ 2; and Lutz to Davenport, 3 Aug. 1908, CBD.

21. M. Smallwood to Zeleny, 28 Aug. 1904, CZ 2. Lutz was already quarreling with Davenport, and Smallwood anticipated "times of great stress."

22. Lutz to Zeleny, 29 May 1905, 3 Dec. 1906, 24 Dec. 1907, CZ 2.

23. Zeleny to Ross Harrison, 25 Jan. 1921, CZ 4.

Figure 2.3   Frank E. Lutz collecting insects, Puerto Rico, 1915. Courtesy of Department of Library Services, American Museum of Natural History, New York (negative 239930).

mutants began to turn up. In July 1908 Lutz also found a dwarf mutant in his cultures.[24] The appearance of definite and clearly heritable morphological characters made *Drosophila* a more appealing organism for experimental heredity. That did not make it better for experimental evolution, however, and experimental evolution was what Lutz really wanted to work on.

In August 1908 Lutz asked Davenport, who was director of Cold Spring Harbor, to underwrite a large project in experimental evolution—not with *Drosophila,* but with *Gryllus.* The idea was to combine his ongoing breeding experiments on short- and long-winged crickets with embryological research (to show how environment affected the development of characters) and, most important, with a full-scale field study of variation in natural populations in Mexico or Cuba. For experiments on evolution, *Gryllus* had obvious advantages over *Drosophila.* Its distinctly dimorphic wing shape was like the Mendelizing traits of domesticated creatures, and its large size made it feasible to do embry-

24. Lutz to Zeleny, 30 Oct. 1908, CZ 2; and Lutz to Davenport, 3 Aug. 1908, CBD.

ological dissection and quantitative biometric work in the field. Drosophilas were just too small and mobile, and all alike.

The combination of lab work and fieldwork was crucial for experimental evolutionists like Lutz or William Tower. It did not make sense to study variation or behavior in the laboratory without knowing that the traits being studied were actually important to the animals' life in the wild. Without fieldwork, studying heredity in the lab was "mere breeding," as Lutz put it. Heredity was of interest only as it pertained to natural variation and evolution. Nature was the real thing, the laboratory an artificial simulacrum. If Lutz could not get money for fieldwork, he told Davenport, he would have to give up *Gryllus* and just do *Drosophila*—"mere breeding." Davenport was not sold on the idea, however, and Lutz left Cold Spring Harbor soon afterward for the American Museum of Natural History, taking his drosophilas with him. Dropping the fruitless work on rates of development, he began selecting pure strains of his new wing mutants to see if they Mendelized. They seemed to, he found, but not as predicted by Mendel's law; so Lutz went on to other bugs and other problems.[25]

In sum, the early breeding experiments with *Drosophila* should not be seen as part of the neo-Mendelian revival. For genetic experiments, domesticated creatures, with their Mendelizing traits, had the clear advantage. *Drosophila* was brought into the lab for studies of variation and adaptation, that is, experimental evolution. In such work its lack of dimorphic traits was no disability. Indeed, it was an advantage for exploring variability and selection. *Drosophila* first occupied the domestic microenvironment for which its natural qualities made it best adapted; namely, that ill-defined region where heredity, embryology, and evolution met.

Drosophila: A Teaching Instrument?

The chief advantage of *Drosophila* initially was one that historians have overlooked: it was an excellent organism for student projects. Drosophilas kept an academic schedule. They were most plentiful in the autumn, when gardens and orchards were filled with rotting fruits and

25. Lutz to Davenport, 3 Aug., 19 Dec. 1908, 17 Feb. 1909; W. L. Tower to Lutz, 8 Sept. 1908; Lutz, "Alternative plan for cricket work," n.d. [Aug.–Sept. 1908]; Lutz, "Report concerning a trip to Mexico and Cuba, May 8 to June 18, 1908"; all in CBD. See also Frank B. Lutz, *Experiments with Drosophila*, CIW publication no. 143 (Washington, 1911).

vegetables, and as they were accustomed to wintering over in human habitations, they happily bred year-round in the warmth of college labs. They were easily replaced and inexpensive to keep, thriving on anything that would support a growth of yeast (bananas quickly emerged as the ideal: not too juicy, cheap, and available year-round). If colonies were killed off by inexperienced or careless students, only a modest investment was lost. When cultures died out in the summer vacation, it was easy to start new cultures when student laborers returned in the fall. The craft skills of *Drosophila* culture were easily learned, and a fly colony could be easily maintained in the interstices of a busy undergraduate schedule. Animal colonies and greenhouse collections were far more exacting of time and skill and were too valuable to entrust to succeeding generations of green apprentices. Invisibility could also be advantageous. Drosophilists would not be called on the carpet, as Castle was in 1903, to reassure President Charles W. Eliot that his hordes of rodents would not escape from their improvised quarters and devour the property of the other occupants of their shared space.[26] Finally, *Drosophila* was convenient for classroom demonstrations of biological principles (professors preferred live material) and for short-term student projects on tropisms, sexual dimorphism, or metamorphosis, for example. Lutz sang the praises of the little fly to biology teachers—"a most excellent laboratory material."[27]

With their low status and undemanding lifestyle, drosophilas were well adapted to the domestic economy of collegiate and university research. Departments of biology were cash poor but rich in one resource: cheap, eager, and renewable student labor. Student scholarships and assistantships were a major subsidy of research, comparable to government subventions of agricultural research stations or foundation endowments of research institutes. Laboratory creatures and their human symbionts followed the money in their own ways. Plant genetics and animal husbandry flourished in agricultural colleges, with their subsidized greenhouses and experimental farms. Small animal geneticists gravitated to research institutes, like the Carnegie Institution's laboratory at Cold Spring Harbor and the Jackson Laboratory at Bar Harbor, whose patrons took pride in large expensive projects that universities could not undertake. As Frank Lutz observed, *Drosophila* breeding, unlike natural history fieldwork, did not require the re-

26. Castle to Jerome Greene, 1 Jan. 1903; and Greene to Castle, 29 Dec. 1902; Charles W. Eliot Papers 133.897, Pusey Library, Harvard University.

27. Frank B. Lutz, "The merits of the fruit fly," *School Science and Math.* 7 (1907): 672–73.

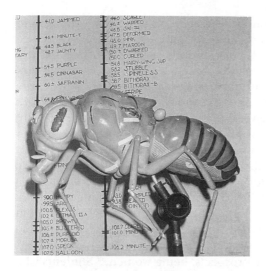

Figure 2.4   Model *Drosophila* with interchangeable mutant eyes, bristles, and wings. Made for H. J. Muller in the late 1940s and used at Indiana University for classroom demonstration. Courtesy of Thomas Kaufman, Indiana University Department of Biology, Bloomington. Photo by Jordan Marché.

sources of a research institution like the Carnegie; it could be done at any college.[28] And it was. Drosophila thus gained a small niche among its more cultivated and prestigious competitors in academic laboratories.

At Harvard almost all the early work on *Drosophila* was done by undergraduates and beginning graduate students. Castle's five-year study of inbreeding was managed by a succession of students, each of whom took charge for one academic season. (Castle helped keep the stocks going through the summer months.) The *Drosophila* project expanded in part because Castle needed such a project for his growing group of students. As he noted, his promotion to assistant professor in 1903 gave him "more opportunity to have student associates in minor research projects."[29] But with the privilege of access to the prime academic resource came the chore of finding doable "minor" projects for beginning students. For that the *Drosophila* project was ready-made. This symbiosis of new generations of flies and students was not without risk, however. Castle observed that his population of drosophilas rose and fell with the seasons, flagging in the fall and recovering in the spring. Probably, as Moenkhaus surmised, he was measuring the learning curve of generations of novice student drosophilists.[30]

28. Lutz to Davenport, 3 Aug. 1908, CBD.

29. Castle, "Memorandum on the beginning of genetic studies" (see n. 16); and Castle et al., "Inbreeding, cross-breeding," pp. 735–36, 742, 747–49.

30. Castle to Davenport, 27 June 1906, CBD; and Moenkhaus, "Inbreeding and selection," p. 131.

There was a similar connection between *Drosophila* culture and teaching at Columbia, where first-year graduate students were expected to do minor projects with various members of the faculty, to get firsthand knowledge of what each was doing. That was how Fernandus Payne came to work on *Drosophila*. Morgan thought Payne should try his hand at breeding work, and *Drosophila* was handy. "With Morgan I will probably try several things," Payne wrote Charles Zeleny, "for I fear some will give no results. One thing I expect to do is to try to rear some aphids and flies in the dark. On the flies I expect to try all sorts of things to see whether I can produce variations."[31] With mice such short-term projects would have been much harder to arrange.

It is striking how many of the people whom Morgan got to work on *Drosophila* in the early years were, like the fly itself, relatively low in academic status. These included his technicians Eleth Cattell, Sabra Tice, and Edith Wallace, who were capable college graduates in biology with distinctly limited career prospects, being women.[32] Students who worked on *Drosophila* also tended to be those who had little to lose by investing in a low-status and uncertain creature. Ann Elizabeth Rawls was a masters candidate (M. A. 1912), and Leopold Quackenbush was a long-term graduate student (M. A. 1901) who was chronically ill, possibly with tuberculosis, and who dropped out and then died a few years later.[33] (It was Quackenbush who Morgan hoped might collect snails in Hawaii while restoring his health through outdoor work.) Morgan even went so far as to recruit Columbia undergraduates, such as Joseph Liff, Alfred Sturtevant, and his part-time bottle washer Calvin Bridges. Ambitious graduate students chose less risky and higher-status projects that were more likely to lead to publications and careers. Payne, for example, chose to do his dissertation with Wilson on spermatogenesis in the toad bug, a decidedly less chancy creature than *Drosophila*.

Quite a number of Morgan's early drosophilists were college teachers who carried out thesis research on the job, over a period of years. Gail Carver was professor of biology at Mercer College in Macon, Georgia, and John Dexter professor of biology at Northland College in Ashland, Wisconsin, when Morgan recruited them to work on *Drosophila;*

31. Payne to Zeleny, 7 Oct. 1907, CZ 2. Payne did similar experiments with Wilson, irradiating tadpoles and other organisms in the hope of making chromosomes segregate in abnormal ways.

32. Wallace, whose beautiful paintings and drawings of *Drosophila* are reproduced to this day, was also epileptic.

33. *Catalogue of Officers and Graduates of Columbia University*, 16th ed. (New York: Columbia University, 1916).

Roscoe Hyde was professor of biology at Indiana State Normal.[34] Clara Lynch did her *Drosophila* work as an instructor at Smith College, and Shelley Safir, while teaching biology in New York City high schools. Mary Stark began her interesting work on a heritable tumor in *Drosophila* while a professor of biology at St. Olaf's College in Northfield, Minnesota. The proliferation of small colleges in the late nineteenth century, especially in middle America, produced a large pool of college-trained, experienced recruits for graduate education. Secure employment and isolation from the centers of experimental biology enabled them to take a chance on a marginal creature. There were few options, and *Drosophila* was an ideal creature for work that had to be fitted into the interstices of busy teaching careers. *Drosophila* thus flourished in the Spartan ecology of the American college system.

Morgan: *Drosophila* and Mutation

Why did Morgan adopt *Drosophila* and press it on his students? What role did he think it might play in experiments on evolution? We need to look more closely at the possibilities for *Drosophila* in specific modes of practice. It is well known that Morgan was hoping that it might be induced to mutate. He did not believe that new species were created by natural selection of minute random variations, as Castle and others did. Selection might preserve a new species, Morgan thought, but only repeated mutation could create one. Inspired by Hugo de Vries's prediction that natural populations entered periodically into episodes of mutation, Morgan began to hope that these "mutating periods" might be observed in organisms other than *Oenothera,* or even induced experimentally, indoors.[35] The Hawaiian land snails that Morgan hoped to

34. J. Dexter, "On coupling of certain sex-linked characters in *Drosophila*," *Biol. Bull.* 23 (1912): 183–94, on p. 183; R. Hyde, "Fertility and sterility in *Drosophila ampelophila*," *J. Exp. Zool.* 17 (1914): 141–71; and G. L. Carver, "Studies on the productivity and fertility of *Drosophila* mutants," *Biol. Bull.* 73 (1937): 214–20 (note the late date).

35. Allen, *Morgan,* pp. 106–25, 144–53; Allen "Introduction of *Drosophila*," pp. 326–27; Garland E. Allen, "Hugo de Vries and the reception of the 'mutation theory,'" *J. Hist. Biol.* 2 (1969): 55–87; Allen, "Thomas Hunt Morgan and the problem of natural selection," *J. Hist. Biol.* 1 (1968): 113–139; Peter J. Bowler, "Hugo de Vries and Thomas Hunt Morgan: The mutation theory and the spirit of Darwinism," *Annals Sci.* 35 (1978): 55–73; Sturtevant, "Reminiscences of T. H. Morgan," 16 Aug. 1967, AHS 3.20; and C. B. Davenport, "The mutation theory in animal evolution," *Science* 24 (1906): 556–57.

study in 1905 were, he surmised, in such a mutating period, and it was shortly after his hopes for a field study collapsed that he turned to experiments in the lab. The first of these were done in the summer of 1906 at Woods Hole.

For these experiments Morgan used insects, "injecting salts, ferments etc. into the pupae in the region of the reproductive cells in the hope of producing changes in the germ cells in the next generation." Why insects? He was probably inspired by Wilson, who in 1905 had begun to survey various species of *Hemiptera* in connection with his study of sex chromosome.[36] But what insects? Not drosophilas: their pupae would have been too small to inject. Morgan must have used large insects, collected from the abundant natural summer stock on Cape Cod. Although he had begun too late in the season to detect results, Morgan saw promise in the method and the material.

Morgan apparently did nothing more with insects until the fall of 1907, when he persuaded Fernandus Payne to try inducing mutations experimentally in *Drosophila*. It is not hard to see why he chose *Drosophila* this time: it was readily available in the city, and whereas small size was a disadvantage in injection experiments with pupae, for work with adult insects large numbers offered a greater probability of detecting a low rate of mutation. Payne subjected his drosophilas to heat, cold, centrifuging, and even the effects of X rays. In the single batch of X-rayed flies he found several flies with wing defects that seemed to be inherited; however, he did not continue the experiment, apparently because the physicists at Columbia would not let him use their radium source. A year later another student centrifuged drosophilas and observed another apparently heritable wing mutant, but Morgan could not be sure that the defect had been caused by the treatment. Morgan tried some experiments himself that fall (1908), raising *Drosophila* on a variety of foods laced with different acids, bases, salts, sugars and other nutrients, and alcohol. The results were tantalizing but inconclusive.[37]

36. Morgan to Driesch, 16 Aug. 1906, THM-APS (quotation); and T. H. Morgan, "Edmund Beecher Wilson, 1856–1939," *Biog. Mem. Nat. Acad. Sci.* 21 (1940): 315–42.

37. Payne to Sturtevant, 16 Oct. 1947, AHS 3.19; Morgan to Driesch, 20 Jan. 1908, THM-APS; and C. B. Bridges and T. H. Morgan, *The Third-Chromosome Group of Mutant Characters of Drosophila melanogaster,* CIW publication no. 327 (Washington, 1923), p. 31. Morgan's experiments are here dated 1909–10. As a beginning student Sturtevant also tried to produce wing mutants using radium. Sturtevant, untitled MS, n.d. [1911?], AHS 16, file Early Manuscript.

All this was not random trial and error; there was method in it. In de Vries's theory, mutating periods were induced by changes in the environment in which species lived. Mutation, according to de Vries and Morgan, was adaptive and responsive. A mutation period might be triggered, for example, if creatures were exposed to circumstances that were unusually favorable for proliferation;[38] or, equally plausibly, when their survival was threatened by adverse conditions. Payne and Morgan's experiments with heat, cold, chemicals, and various foods were clearly meant to explore these two possibilities. Yet another possibility was that creatures entered into a mutating period when subjected to intense selection.[39] It was the last idea, I believe, that inspired Morgan's experiments on selection and inbreeding, in the course of which the white-eyed mutant unexpectedly turned up.

In the fall of 1909 Morgan began to experiment on the variable trident-shaped pattern on the thorax of *Drosophila*, selecting and inbreeding stocks in which the trident marking was least and most pronounced. This kind of "pure-line" selection was the commonest variety of experimental evolution at the time. Neo-Darwinians like Castle used it to model in the lab what they believed to occur as species evolved in nature. Anti-Darwinians like Wilhelm Johannsen used the same experiments to show that selection of minute variations could not result in permanent change. So what did Morgan hope to show with his selection of the trident marking? Certainly he was not emulating Castle; nor, I think, was he recapitulating Johannsen's classic work.

The timing of the trident experiments, coming not long after Lutz's discovery of wing-vein and dwarf mutants, suggests that Morgan was trying to reproduce Lutz's work and to explore the possibility, which Lutz had not, that the experimental process of selection had itself produced mutations.[40] This is supposition, but it is consistent with Morgan's customary working habits, and we do know that he and Lutz were exchanging *Drosophila* stocks and lore at the time. In the summer of 1908, just after his wing mutants turned up, Lutz went to Woods Hole especially to see Morgan about them: "I am sort of run out of

38. D. T. MacDougall, "Discontinuous variation and the origin of species," *Science* 21 (1905): 540–43.

39. T. H. Morgan, *Evolution and Adaptation* (New York: Macmillan, 1903), pp. 287–93, 461–62; and Morgan, "For Darwin," *Pop. Sci. Mon.* 74 (1909): 367–80.

40. Morgan later described these experiments as a "side issue" in his line of experiments on chemical and physical mutagenesis but did not elaborate: Bridges and Morgan, *Third-Chromosome Group*, p. 31 and fig. 4.

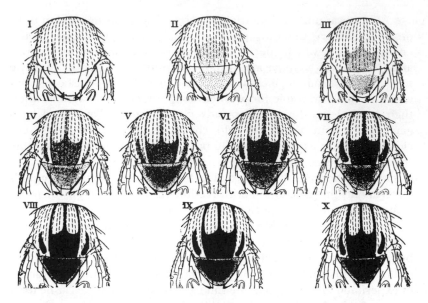

Figure 2.5   Variations in the trident pattern of *with* flies, in which T. H. Morgan discovered his first mutant in January 1910. From Bridges and Morgan, *Third-Chromosome Group,* on p. 31.

inspiration," he confided to Zeleny, "and I know of no better place to get some more.[41]

But exactly what kind of mutational event did Morgan expect to see? It has usually been assumed that he was looking for what he actually found: a visible, striking change like the white-eyed mutant. But Morgan could not have been expecting to produce de Vriesian macromutations, because he did not believe that such large mutations gave rise to new species in nature. They were too rare and too extreme. The raw materials of evolution, in Morgan's view, were definite mutations within the range of normal variation of a species or variety. If one of these small but definite mutations could induce another to occur in the same direction, the process could, he thought, lead to the rapid evolution of new species.[42] These mutations would appear not as visible

41. Lutz to Zeleny, 31 July 1908, CZ; see 2 Sept., 30 Oct. 1908. Lutz sought Morgan's approval of his *Drosophila* paper before publishing it and supplied him with stocks: Sturtevant to J. W. Wilson, 22 Sept. 1965; and Wilson to Sturtevant, 20 Sept. 1965; AHS 3.19.

42. Morgan, *Evolution and Adaptation,* pp. 101–104, 261–97, 452–64; Morgan, "The origin of species through selection contrasted with their origin

characters but as abrupt steps in the selection process. Apparently stable "pure" lines of high and low trident would suddenly be capable of further and more extreme modification by selection. That, I think, is what Morgan expected a de Vriesian mutating period to look like if it could indeed be induced in the laboratory by intense artificial selection.

The selection experiment with trident was different from Morgan's earlier experiments with chemically and physically induced mutation, in which he may well have been hoping to produce large, visible mutants. Morgan appears to have designed the trident experiment to simulate more closely the actual process of evolutionary change in the wild. He was hoping to invent a new mode of experimental evolution that was different from those of de Vries, Castle, and Cuenot precisely in dealing with the kind of variations on which, he suspected, selection acted in nature. Morgan disparaged neo-Darwinian experiments with fluctuating characters as mere "meddling" with an organism, upsetting their natural physiological balance in a way that could never occur in nature.[43] To him, selection experiments like Castle's were laboratory exercises that had nothing to do with evolution in the wild. Nor did the experiments of Neo-Mendelians like Cuenot, in his view; they might illuminate the process of hereditary transmission but were irrelevant to evolution because they dealt only with characters artificially altered by domestication. For Morgan, a true experimental evolution dealt with the kind of variations from which new species were actually created in nature: that is, mutations that were discontinuous but within the normal range of variation.[44] In other words, Morgan was using the selection method of the neo-Darwinians to detect mutations too small to show up in de Vries's approach.

Experience did not immediately conform to expectations. By inbreeding lines with light and dark trident patterns, Morgan quickly produced strains that selection could not further change. But then for months nothing happened. Morgan grew discouraged. Ross Harrison recalled visiting Morgan's lab in the first days of January 1910: "'There's two years' work wasted,' he [Morgan] exclaimed, waving his hand at rows of bottles on shelves. 'I've been breeding these flies for all that time and have got nothing out of it.'" But just a few days later,

---

through the appearance of definite variations," *Pop. Sci. Mon.* 67 (1905): 54–65; and Morgan, "For Darwin," pp. 375–77, 379–80.

43. Morgan, "Origin of species," pp. 63–64.

44. Morgan, *Evolution and Adaptation*, p. 286.

Morgan observed a few flies with a darker pattern than any he had seen before. Inbreeding quickly produced a mutant strain, *with,* which had a distribution of pigmentation that was distinctly higher than the wild type, and which further selection did not alter. It was just what Morgan expected the process of speciation by mutation would look like! Visiting his wife and newborn daughter in the hospital in the second week of January, Morgan could talk of nothing but his new mutant. No wonder: Morgan thought he was witnessing the beginning of a mutating period—evolution induced by strenuous selection in the laboratory, where all could witness it![45] A second mutant, *superwith,* appeared in November 1910, thirty generations later.

Morgan's various efforts to induce mutation by altering the physical environment and by selection were in my view preparations for a major project in experimental evolution, parallel to regeneration, sex determination, and Mendelizing. The pattern of his behavior is unmistakable. He wrote his 1903 book and subsequent articles in 1905 and 1909 to assimilate current work on adaptation and evolution and to stake out a personal position on key issues. He did experiments on *Drosophila* and other insects to get hands-on experience of the ways that experimental evidence about variation was produced. This was just how he had broken into the neo-Mendelian game, criticizing Cuenot and getting a feel for mouse breeding. By 1909 he had a distinctive view of variation that served as a practical guide for breeding experiments, and he thought that *Drosophila* might be the ideal organism for experiments on evolution, just as mice were for Mendelizing and *Phylloxera* was for sex determination.

Morgan chose *Drosophila* for experimental evolution, rather than some domesticated plant or animal, precisely because it was *not* domesticated. He had long believed that experiments on evolution would have to be done with wild creatures. "Until we know more about the results when wild varieties are crossed with their wild species or with other varieties," he wrote in 1909, "we can not safely apply Mendel's

45. Bridges and Morgan, *Third-Chromosome Group,* pp. 31–37; R. G. Harrison, "Embryology and its relations," *Science* 85 (1937): 369–74, on p. 370; Allen, *Morgan,* pp. 148–53; Morgan, "Hybridization in a mutating period in *Drosophila,*" *Proc. Soc. Exp. Med.* 7 (1910): 160–61, and Elof A. Carlson, "The *Drosophila* group: The transition from the Mendelian unit to the individual gene," *J. Hist. Biol.* 7 (1974): 31–48. Morgan's daughter was born on 5 Jan. 1910.

laws to the process of evolution."[46] It is also clear why Morgan did not use Lutz's domesticated stocks of *Drosophila* but instead collected fresh material from nature. If Morgan believed that Lutz's inbred stocks had already entered into a mutating period, he would have wanted to avoid such stocks in experiments designed to see if artificial selection could induce such an event. (Morgan also insisted that Payne use a wild stock for his selection experiments, not one from Lutz, doubtless for the same reason.)[47] *Drosophila's* natural wildness and lack of Mendelizing characters, which so limited its use for genetic experiments, made it the organism of choice for Morgan's new program in experimental evolution. Morgan meant *Drosophila* not to replace mice but to establish a new niche and a new line of experiment in the ecology of his lab.

Drosophila Takes Over

The appearance of *with* in January 1910 was just the harbinger, as it turned out, of a gathering flood of mutants: *olive* body color and *speck* wing axil in March, *beaded* wing and another *olive* in May, along with the famous *white* eye-color mutant. *Rudimentary* wing appeared in June, *pink* eye in July, *miniature* and *truncate* wings in August, and at least six more by the end of the year. It appeared to Morgan that his selection experiment had indeed inaugurated a de Vriesian mutating period in his flies. The character of these mutants, however, was not quite what the theory predicted. Only the *olive* mutants were like *with*, definite but within the range of normal variation. The others were more extreme, more like de Vries's *Oenothera* mutations of Mendelizing sports. Although they did not display precise Mendelian ratios, these mutants did segregate. And *white* eye, to Morgan's astonishment, was sex-linked. That is, it appeared only in males, which had therefore to have a single sex chromosome. This freshet of extreme mutations was not what Morgan had expected of a mutating period, and it entirely upset any plans for a project in experimental evolution using *Drosophila*.

Indeed, the freshet of mutants in *Drosophila* completely upset the hierarchy of experimental lines and organisms in Morgan's lab. With

46. Morgan, "Breeding experiments with rats," *Amer. Nat.* 43 (1909): 182–85, on p. 183; and Morgan, *Evolution and Adaptation*, p. 278. Morgan crossed wild black and Alexandrian rats in 1908–9.

47. Bridges and Morgan, *Third-Chromosome Group*, p. 225; and Payne to Sturtevant, 16 Oct. 1947, AHS 3.19.

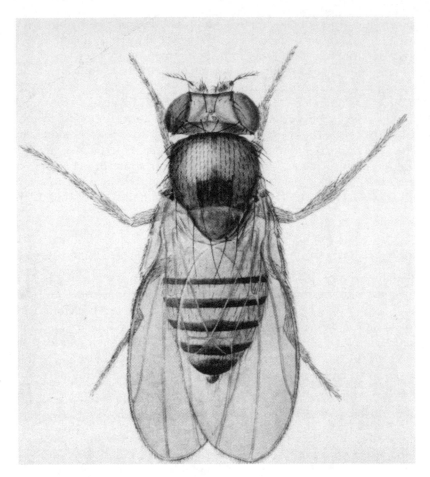

Figure 2.6 *Speck olive,* an early variable mutant, hand drawn and colored by Morgan's assistant and fly artist, Edith M. Wallace. From Bridges and Morgan, "Second Chromosome Group," plate 5, at p. 128.

the appearance of *white* and other Mendelizing mutants, *Drosophila* broke out of its restricted domain of experimental evolution and began to take over Morgan's other experimental lines. The sex-linked character of *white* made it an ideal organism for work on sex determination, better even than *Phylloxera,* which was soon set aside for studies of sex chromosomes and sex determination in *Drosophila*. Similarly, the fact that *white* and other mutants displayed more or less Mendelian inheritance made *Drosophila* the organism of choice for experimental heredity, better than mice or rats with their limited number of characters. Mice were soon displaced from their favored niche in the Columbia lab. In short, the proliferation of mutations in *Drosophila* altered the domestic ecology of experimental organisms and disciplines. The fly began to take over Morgan's entire operation, displacing the standard organisms in his other lines of work.[48] Morgan's mouse colony had been little used for some years before it was finally destroyed in a fire in 1914. *Drosophila* had taken over its role in experimental heredity as effectively as it had displaced *Phylloxera* from experiments on sex determination. No creature could compete with the little fly.

Experimental evolution, ironically, was the one line that did not survive *Drosophila*'s takeover of Morgan's lab. As it gradually became clear that the flood of mutants resulted simply from scaled-up production and not from a de Vriesian mutating period, selection experiments ceased to have any relevance to the mechanism of evolution. Besides, Morgan was not one to miss an experimental windfall, and the Mendelian analysis of mutants like *white* was the kind of opportunity that comes seldom in a lifetime. Morgan did continue selection experiments on *with* and similar mutants for a year or so, but his gradual abandonment of this line of work for neo-Mendelian experimental heredity marks the transition from experimental evolution to modern genetics.

Morgan's last brief fling with experimental evolution began in November 1912 with the appearance of *streak*, a slight and highly variable thorax-pattern mutant. Bridges and Sturtevant would have discarded it as useless for mapping, but since it had turned up in a stock that had never been subjected to selection, Morgan saw a chance to prove what everyone now believed, that selection did *not* induce mutations of the *with* type. He worked on *streak* for six months, "not very vigorously" and to little effect. The next such mutant that appeared, *trefoil* (November 1913) was discarded forthwith, and in June 1914 all the *with* stocks were thrown out and soon forgotten.[49] It is striking how completely this

48. T. H. Morgan, "Chromosomes and heredity," *Amer. Nat.* 44 (1910): 449–96; Morgan, "Sex-limited inheritance in *Drosophila*," *Science* 32 (1910): 120–22; Morgan, "The application of the conception of pure lines to sex-limited inheritance and to sexual dimorphism," *Amer. Nat.* 45 (1911): 654–78, on pp. 674–78 (this talk was delivered in December 1910); and Allen, *Morgan*, pp. 137, 148–52.

49. Bridges and Morgan, *Third-Chromosome Group*, pp. 31, 36–44, and C. B. Bridges and T. H. Morgan, "The Second-Chromosome Group of Mutant Characters," in *Contributions to the Genetics of Drosophila melanogaster*, CIW publication no. 278 (Washington, 1919), pp. 125–342, on pp. 136–40, 222, 244–45.

crucial stage in the invention of genetics was wiped from the drosophil-ists' collective memory. When Lillian Morgan recalled her husband's visit to the hospital in January 1910, she remembered him talking about the *white* mutant. It must have been *with*, however, since *white* only turned up in May!

The invasion of *Drosophila* mutants thus turned Morgan's orderly hierarchy of experimental organisms and disciplines topsy-turvy. *Drosophila* took over its new ecological niche, displacing established organisms and imposing a new order on the organization of experimental work. It would change forever the natural history of experimental biological laboratories.

Drosophila: A Breeder Reactor

The cause of this fundamental change in experimental practice was the appearance of large numbers of mutants, and that phenomenon was in turn caused by the scaling up of Morgan's breeding experiments. The more flies that were bred and inspected, the greater was the probability that mutants would turn up. At some threshold of scale (perhaps five or ten thousand flies) a selection experiment would have contained enough spontaneous mutants for one to be noticed. Doubtless many mutants were passed over unnoticed, but once the first was observed, others were more likely to be picked up—hence the apparent suddenness of their appearance.[50] But if it was just a phenomenon of scale, why then did Castle not also observe mutants, since he was working on a scale at least as large as Morgan and Lutz? The answer is simple: Castle was working on fertility, and to reduce the burden of routine he counted only pupae. As a result he worked only with a small number of adult flies and so never had a chance to spot mutants.[51]

Mutants did not turn up in all experiments but only in certain specific kinds; the causal connections are quite specific. Mutants could only have turned up after Morgan and Lutz began to work on experimental evolution, because only these experiments were big enough to cross the statistical threshold. Morgan's early efforts to induce muta-

50. It may also be that the crossing of inbred and wild stocks increased the natural rate of mutation through an exchange of extrachromosomal genetic elements, or *transposons*. I am indebted for this idea to Edward B. Lewis of Caltech. See James F. Crow and William F. Dove, "The genesis of dysgenesis," *Genetics* 120 (1988): 315–18.

51. Castle, "Memorandum on the beginning of genetic studies" (see n. 16).

tion by heat, cold, or X rays were doubtless done with cultures of a few hundred flies—not big enough. Scaling up was crucial: it was only in the inbreeding experiments with the trident pattern, which involved many generations of mass cultures, that mutants were bound sooner or later to be spotted. Mendelizing experiments might have been big enough to cross the statistical threshold, but *Drosophila* was not well suited to such experiments, as we have seen. That is why *Drosophila* first found a habitable and expandable ecological niche in experiments on evolution. It was only in this line of work that its potential for mutation could be revealed.

Once the threshold of scaling up was crossed, the production of new mutants and new genetic knowledge fed on itself. The more mutants turned up, the more crosses had to be done to work them up. The more crosses were done, the more mutants turned up. The process was autocatalytic, a chain reaction. *Drosophila* became, in effect, a biological breeder reactor, creating more material for new breeding experiments than was consumed in the process.

This autocatalytic property of *Drosophila* breeding was specific to experiments on neo-Mendelian heredity. Striking mutations like *white* were not, strictly speaking, material for experimental evolution, at least as Morgan conceived it. Experiments on evolution required slight mutants like *with* or *superwith,* but these were rare because they could not be identified by simple inspection. Most of the mutants that turned up were like *white.* Thus there was no feedback loop, no autocatalytic potential to the selection work—it was no breeder reactor. Neo-Mendelian breeding experiments were different, because most of the mutations that turned up were raw material for further experiments of the same kind. A feedback loop began to operate which resulted in a continual scaling up of the size and scope of breeding experiments. Morgan was compelled to work up each new mutant and integrate it into the genetic formulas for eye color, wing shape, and so on (more on that in the next chapter). Neo-Mendelian experiments were the first mode of experimental breeding that produced, as a by-product, more material for experiments of the same kind than they consumed. This mode of experiment was the first true breeder reactor.

The productivity of the *Drosophila* breeding system was relentless. In November 1910 Morgan wrote his friend Hans Driesch that he had been "overwhelmed with work" since the summer. "It is wonderful material. They breed all the year round and give a new generation every twelve days." In March 1911 he was becoming desperate. "I am beginning to realize that I should have prepared for a large campaign and be[en] better organized," he wrote Charles Davenport, "but who could have foreseen such a deluge. With vicarious help I have passed one acute stage only I fear to pass on to another. With what help I can muster I hope to weather the storm." But there was no let up. In January 1912 he was again "head over ears in my flies," swamped with more mutants than he had time to work up."[52] Even after the productivity of *Drosophila* was well known, mutant hunters were still taken aback by the experience of the breeder reactor. Charles Metz was an old hand when he began systematically to domesticate some new wild species in Florida in 1919, but even he was taken aback: "I don't know whether it is the climate or whether my eyes are better than they used to be," he wrote Davenport, "but the way I am turning out mutants and 'possible mutants' beats any experience I ever had before. If things keep on piling up I will be swamped in another month and will have to call for help." It was, he noted later, "almost an epidemic."[53]

"Epidemics" of mutation regularly struck the Morgan laboratory in the early years. In these episodes, the appearance of a new mutant seemed to trigger the appearance of more like it. An epidemic of *purple* eye color occurred in the first half of 1912, when no fewer than fourteen *purple* mutants turned up. It happened again with *jaunty* wing, when six mutants turned up in just three months. Similar epidemics appeared later with *vermilion, cut,* and *notch.* The Morgan group began speaking of a "mutating period," perhaps by then as a joke on themselves.[54] These epidemics were not de Vriesian mutating periods, of course, but a property of the new mode of practice that was created when *Drosophila* began to be used for large-scale experimental breeding.

Epidemics of mutants were in part the result of imperfect laboratory technique and enthusiasm. Many of the mutants in these epidemics turned out to be duplicates, resulting from accidental contamination of breeding stocks, or simple extreme forms of natural variation in the wild type—phantoms in the minds of mutant hunters. Nevertheless, genuine new mutants kept appearing at an accelerating rate: at least ten in 1911 and two dozen in 1912. By 1915 so many mutant stocks were being cultivated that many of the less useful ones had to be

52. Morgan to Davenport, 14 Mar. 1911, CBD; and Morgan to Driesch, 23 Nov. 1910, 1 Jan. 1912, THM-APS.

53. Metz to Davenport, 13 Apr. 1919, 13 July 1922, CBD; see also 24 June 1919.

54. Bridges and Morgan, "Second-Chromosome Group," pp. 161–62, 178–79.

Figure 2.7   T. H. Morgan at work in the Columbia fly room, circa 1915–20. Courtesy of American Philosophical Society Library, Columbia University Department of Biology Photograph Collection.

discarded; there were just too many to keep. Such was the profligate fecundity of *Drosophila* in its new domestic ecosystem.

Morgan and his drosophilists did indeed discover a "mutating period," but it had nothing to do with evolution in nature. It was an aspect of the natural history of experimental laboratories and experimental practices. The awesome productivity of *Drosophila* was not simply a physiological property of the organism itself, though it was partly that. Rather, it was a property of a system of experimental mass production and of the symbiotic relationship between flies and fly people. The capacity to produce mutants was the property of a creature with a fast life cycle and large families when it was brought indoors into a laboratory and integrated into a system for doing large-scale breeding experiments. The breeder reactor was the whole system of production: the constructed mutant stocks and standard recipes, drosophilists' research agendas and ways of life. This symbiotic relationship would in time transform *Drosophila* physically into a new domesticated creature, one that did not exist in nature and that could only have been created in the peculiar ecology of a genetics laboratory. So, too, did it transform the fly people into a new variety of experimental biologist, with distinctive repertoires of work and a distinctive culture of production.

Conclusion

Many creatures have over the years crossed the threshold between nature and experimental laboratories, but few have been such successful colonizers of this "second" nature as *Drosophila*. Few symbiotic relationships between experimenters and their laboratory creatures have been so fruitful for so long. Few standard laboratory creatures have evolved into such a vast and diverse extended family. In its new ecological space *Drosophila* became a breeder nonpareil of new experimental material and practices. It altered the rules of the games that experimental biologists played and changed the natural history of their domestic workplaces.

It was only *after* it had a foothold in the laboratory, however, that *Drosophila*'s potential as a breeder reactor was revealed, and the little fly got across the threshold in the first place because it was adaptable generally to academic life. *D. melanogaster*'s natural history in orchards and cider mills, I have argued, matched the natural history of academic departments. Its seasonal demography matched the seasonal cycle of student projects and student labor; cheap and lowly, *Drosophila* was not shut out by the poverty of academic research funding. But once inside, the little vinegar fly revealed to experimenters what it was good for. Like many other invaders of new ecological territories—gypsy moths, English sparrows, Russian thistles—Drosophila was introduced innocently and then unexpectedly took over, displacing indigenous inhabitants by outproducing them. Only by entering through the back door could such an inconspicuous and unlikely creature have entered into competition with mice, primroses, and other well-entrenched and well-adapted inhabitants of experimental labs.

Drosophila took over Morgan's laboratory in a way that was neither intended nor predictable. Indeed, *Drosophila* helped to create an environment in which it could thrive by setting off a rapid succession in the modes of experiment in which Morgan was engaged. The lowly fly's short life cycle and large families drew Morgan into using it in a new variety of experimental evolution that required breeding and selection on a very large scale. As a result of this scaling up, *Drosophila* revealed its capacity to produce large numbers of well-defined mutations, and that capacity impelled Morgan to abandon experimental evolution for the neo-Mendelian variety of experimental heredity, for which the fly, as an undomesticated creature, had seemed initially ill suited. In this mode of experimental life *Drosophila* revealed the further novel and remarkable capacity, observed in no other laboratory creature, to produce more material for experimentation than it consumed—the biological breeder reactor. That capacity would in turn force Morgan to abandon the neo-Mendelian mode of heredity for a new mode of genetic mapping, as we will see in the next chapter. *Drosophila* drove this succession of experimental practices in ways that no one could have foreseen.

It is a remarkable fact that *D. melanogaster,* the original colonizer of Morgan's lab, was never displaced as the standard fly. Many other species of *Drosophila* eventually joined it indoors and proved better suited to certain specialized uses. But none displaced *melanogaster* as the standard instrument of experimental genetics. In part this may have been what evolutionary biologists call a *founder effect.* The more elaborate the machinery of experimental production built around a species, the more costly it would be to replace it with another species. But it was not just that. *Melanogaster* was the best adapted to the conditions of experimental production: vigorous, fertile, prolific, easy to manipulate, tolerant, and hardy. It was not a fussy feeder and was quick to recover from the recurring disasters of a genetics lab, such as heat or cold waves, epidemics of mites, and neglect. More specialized species of *Drosophila,* which fed fastidiously on fungus or pollen, proved almost impossible to cultivate in the laboratory because they had evolved to depend on special local habitats. Similarly, some wild tropical species that shunned humankind were highly susceptible to epidemics when in captivity. *Melanogaster* was ideal for life in the lab.

Was it an accident that *melanogaster* was both the first *Drosophila* to enter the laboratory and the best suited for life there? Perhaps not entirely. *Melanogaster* had, after all, been cohabiting with and adapting to humankind for some thousands of years, following agricultural peoples from South Asia around the world. The very traits that made it a good hitchhiker were precisely the same traits that made it a good standard laboratory animal. Because it was hardy and not fastidious about what or where it fed, it was more likely than overspecialized local species both to turn up in drosophilists' traps and to flourish in overcrowded laboratory jars. It was relatively tolerant of heat and cold from following humankind into inclement regions and wintering over in chilly fruit stores and cellars. It was already an indoors creature, accustomed to human habitations and able to brave hostile urban environments. Cosmopolitan species, in short, were preadapted to laboratory life, and *D. melanogaster* was the cosmopolitan fly par excellence. It had been shaped by natural selection to live in a commensal relationship with humankind. Like *Homo sapiens, D. melanogaster* thrived because it was an opportunist. There is less difference than we might imagine between laboratories and the places, in and out of doors, where *melano-*

gaster had long flourished as a hanger-on of humankind. The forces that made it such a successful camp follower of the most mobile and cosmopolitan of primates also suited it well to a symbiotic relation with that special variety of humankind who inhabited experimental laboratories.

THREE

Constructing *Drosophila*

MORGAN'S DISCOVERY OF THE AUTOCATALYTIC PROPERTY of large-scale breeding—the breeder reactor—was a turning point in Drosophila's natural history. Soon the little fly was being redesigned and reconstructed into a new kind of standard laboratory instrument, a living analogue of microscopes, galvanometers, or analytical reagents. Laboratory drosophilas differed physically from their wild ancestors and had a different mode of life. They were not a different species: wild flies could and did interbreed with their domesticated cousins (to the consternation of drosophilists). They were different varieties, like garden plants or domesticated animals. "Standard" drosophilas were constructed from stocks that produced recombination data conforming most closely to Mendelian theory. Their chromosomes were a bricolage, artfully put together from useful pieces of chromosomes of various mutant stocks and cleaned by selective inbreeding of the genetic noise that messed up genetic mapping. The purposes and key concepts of mapping were thus built physically into domesticated drosophilas: what better indication of their artifactual nature?

Ecologically, too, laboratory drosophilas differed from their wild cousins. In the lab they no longer confined their active life to the cool quiet hours around dawn and dusk but took to feeding and breeding around the clock. They had distinctive modes of dispersal, not by wind and banana boat but in cozy, protected mailing tubes. Forms thrived in genetics labs that would have perished in the wild without leaving progeny. The special selective forces of experimental labs favored extreme mutant forms, even if they were marginally viable, and at the same time suppressed hidden genetic variability. In nature it was just the opposite: there, pressures of seasonal changes and competition pruned less-viable forms but favored variability. In the wild there are no standard flies, but in the lab standard drosophilas had a selective advantage in their relation with the fly people, who lovingly bred, protected, and dispersed them around the world.

The construction of a standard *Drosophila* was not an inevitable

53

consequence of a domestic mode of life. Rather, it was a consequence of the fly's involvement in genetic mapping, one specific mode of experimental heredity. Neo-Mendelian breeding experiments had no need for a standard fly, because this mode of practice was not quantitative. Genetic mapping, in contrast, was quantitative, and thus required a standard mapping technology and a constructed, standard fly. The difference between these two modes of experimental heredity can be seen in their visual representations. The products of neo-Mendelian experiments were formulas representing the combinations of genetic factors that, either by their presence of their absence, together produced a morphological trait. Assembling such formulas required no quantitative work. Genetic mapping, in contrast, produced maps representing the linear arrangement of mutant genes along the chromosomes (figure 3.1). This representation was essentially quantitative: it represented not just the order of genes but the distances between them. The data for these maps could not be produced with just any mutants. Mappers had to create standard *Drosophila* stocks specifically adapted to the peculiar requirements of quantitative measurement. Genetic maps are the blueprints of the standard fly and our starting point in following the construction process.

The story of how the genetic map was invented—the founding myth of modern genetics—is usually told as a story of conceptual change. First Morgan gave up the idea that genetic factors were present or absent and adopted the view that genes existed in different forms, or *alleles,* arranged in definite linear orders along the chromosomes. He then invented the idea of physical "crossing-over" to explain how genes linked together on the same chromosome did nevertheless segregate, and that physical model led him to the insight that the rate of crossing-over between two linked genes should be proportional to the physical distance between them. Alfred Sturtevant, realizing that distances should be additive, went home and constructed the first primitive map of the X chromosome of *D. melanogaster.*[1]

This foundation story is not wrong, but it does fail to show how conceptual changes were driven by the practical imperatives of doing experiments. It also makes the choice between competing modes of

1. A. H. Sturtevant, undated lecture, AHS 3.20; Sturtevant, *A History of Genetics* (New York: Harper & Row, 1965), p. 47; Sturtevant, "The linear arrangement of six sex-linked factors in *Drosophila,* as shown by their mode of association," *J. Exp. Zool.* 14 (1913): 43–59; Garland E. Allen, *Thomas Hunt Morgan: The Man and His Science* (Princeton: Princeton University Press, 1978), p. 167; and Elof A. Carlson, *Genes, Radiation, and Society: The Life and Work of H. J. Muller* (Ithaca, N.Y.: Cornell University Press, 1981), pp. 66–67.

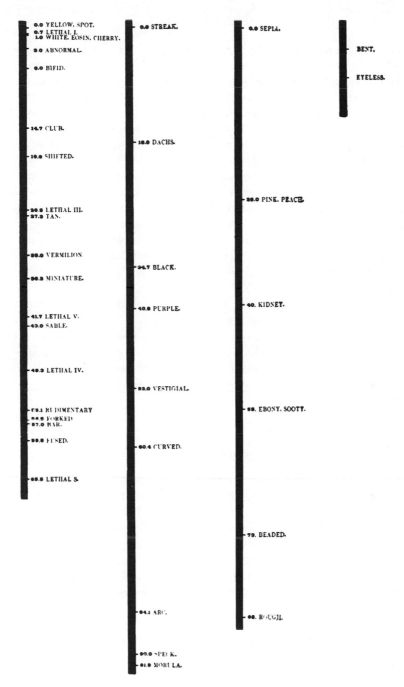

Figure 3.1 Chromosome maps, 1915. From Morgan et al., *Mechanism of Mendelian Heredity*, p. ii.

experiment seem much more rational than it actually was. Viewed as practice, the choice between presence-absence formulas and chromosome maps was first and foremost a choice between different methods of classifying mutants and managing an accumulating mass of experimental data that was rapidly getting out of hand. It was not theoretical preferences that impelled the redesigning and construction of *Drosophila* but practical, mundane problems of managing the deluge of new mutants that the fly spewed forth when it began to be used in large-scale breeding experiments. The neo-Mendelian system of classifying mutants had been designed for experimental organisms with just a few mutant forms, like mice or fowl; with *Drosophila* it simply became impossible to do experiments in the old way. The fly people were driven to invent a system of classifying and managing data that was indefinitely expandable. Hence the genetic map. And hence the standard fly: the one could not exist without the other.

The process of constructing *Drosophila* began not in 1910, when the first mutants appeared, but in early 1912, when genetic mapping began to overtake neo-Mendelian experiments as the central activity of Morgan's lab. We need therefore to begin by taking a closer look at the material culture and practices of these two varieties of experimental heredity and tracing the drosophilists' path from one mode to the other.

Designs: Organ Series and Linkage Groups

The neo-Mendelian system of classifying genetic factors into organ group systems—eye color, wing shape, body color, thorax pattern, and so on—embodied the basic purpose of neo-Mendelian heredity, which was to determine how many genetic factors were involved in the formation of each morphological feature. Genetic crosses were designed to synthesize all possible combinations of present or absent factors. The end products of these experiments were families of genetic formulas, one for each known mutant form. Take the eye-color series in *Drosophila,* for example. By mid 1911 Morgan had five mutant types: *red,* or wild type; *white, pink, vermilion,* and *orange* (which turned out to be a double recessive). To account for the production of these five colors, Morgan postulated the existence of a color determiner, C, and three modifying factors, *red, pink,* and *orange,* symbolized by uppercase R, P, and O when present and, when absent, by lowercase r, p, and o. It was supposed that present factors contributed different pigments to the mixture that made up wild type and that various mutant eye colors were formed by the loss of one or more factors. Thus the red or wild type was *RPO* (*C* was implied), pink was *rPO* (loss of red), orange was *rpO* (loss of red and pink), and vermilion was *RpO* (loss of pink). White eye was produced by loss of *C*, the color determiner.[2] Similar series could be constructed for the families of wing shapes, body colors, bristle patterns, and so on.

This system of classifying mutants into organ groups was fundamentally a physiological and developmental one. One studied the inheritance of factors in order to understand how morphological characters developed. That was how Morgan put it when he tried to explain to the embryologist Hans Driesch why the work with *Drosophila* was not as different as he might think from his own: "You should be [interested]," he exhorted his skeptical friend, "for it turns very fundamentally on the problem of development."[3] Neo-Mendelian heredity, with its organ groups and quasi-chemical elements and compounds, was deeply rooted in a mode of experimental biology in which the boundaries between heredity, development, and evolution were not yet well defined.

Classifying mutants into linkage groups, unlike the organ series system, was structural and spatial. Factors were grouped into families not on the basis of similar developmental functions but simply on the basis of location. Mutant types were symbolized not by different configurations of factors but by single genes: pink eye color was associated with a pink gene, *p*; white eye with a white gene, *w*; purple with the purple gene, *pr*; and so on.[4] This new system of classifying and naming assumed that the purpose of experimental heredity was to determine physical location and not developmental relationships.

The physical manipulations of genetic mapping were much the same as in neo-Mendelian experiments. Different mutant stocks were still crossbred to produce various classes of recombinants. The design of crosses, however, was quite different. Gene mappers never crossed flies with genes in different linkage groups: there would have been no point. In the old system of the organ series it would have been just as nonsensical to cross mutations that affected different organs, just be-

2. T. H. Morgan, "An attempt to analyze the constitution of the chromosomes on the basis of sex-limited inheritance in Drosophila," *J. Exp. Zool.* 11 (1911): 365–410, on p. 367.

3. T. H. Morgan to H. Driesch, 1 Jan. 1912, THM-APS.

4. In this new notation uppercase letters signified dominant genes and lowercase letters recessive genes.

cause they happened to be located near one another. Another fundamental difference was that mappers counted the number of flies in various classes of recombinants. That put a premium on accurate counts and reproducibility. In neo-Mendelian experiments there was no counting: different recombinants were identified and selected out for further crosses.

Like all classification systems—of chemical elements, physical particles, geological strata, or biological species—the linkage-group system directed experimenters toward certain lines of work and away from others. Neo-Mendelian formulas for eye color and other organs constituted the design of an experimental instrument whose purpose was to reveal the genetic causes of morphological development. Location was secondary. Chromosome maps, in contrast, constituted the design of an experimental instrument for locating genes and studying chromosomal mechanics. Developmental relations were set aside. Formula and map are the visual representations, blueprint and graphic summary, of different instruments and alternative modes of experimental production.

Choosing a System

What caused the organ-series system to collapse was the unexpected deluge of new mutants that were produced by scaled-up Mendelian crosses—the breeder reactor. As Morgan recalled, it simply became impossible to do genetic experiments with *Drosophila* using neo-Mendelian formulas.[5] When a new mutant was discovered, for example, it was impossible to give a name to the factor whose loss was presumed to have caused the new form to appear. Indeed, it was necessary to rename all the factors presumed to be involved (for reasons too complicated to explain here). Worse yet, the entire series of formulas in an organ series had to be completely reconstructed every time a new mutant appeared, because the formula for each mutant character had to include all the factors in the series, whether present or absent. This system of classification worked well with creatures that did not produce new mutants, but with *Drosophila* in its breeder reactor mode it quickly became totally unusable. A final problem was that the system's mathematical permutations and combinations of factors predicted the exis-

5. T. H. Morgan, "Factors and unit characters" in Mendelian heredity," *Amer. Nat.* 47 (1913): 5–16, on pp. 6–10, 12–16; see esp. p. 14.

tence of phantom mutant types that should have arisen through re-combination but in fact did not, even when crosses were done on a gargantuan scale.

The failure of the neo-Mendelian system of classification was expe-rienced in part as increasing difficulties in daily work—frustration in routine matters like naming and wasted effort in continually rebuilding basic tools. It was also experienced as a loss of credibility in the commu-nity of experimental biologists. Having continually to change fac-tor formulas was potentially an acute embarrassment, since neo-Mendelians already had a bad reputation for constantly changing their minds. Morgan himself mistrusted neo-Mendelian experiments for just that reason and was very sensitive to the psychosocial dangers of repeat-edly revising what he had published as definitive. Trust is a crucial good for experimental scientists, and it is acquired by producing results that others can use confidently and with assurance that the knowledge will not be continually disavowed. Genetic formulas, like the structural for-mulas of organic chemists, would have to approach some stable, defin-itive form if the entire enterprise were not to be discredited.

It was the change in experimental organisms, basically, that upset the neo-Mendelian applecart. The old rules for naming and classifying factors were designed for creatures like mice or fowl, which had at most three or four different forms in each group and which did not continu-ally produce new ones. The stability of the organ-series classification thus depended on the fixed genetic character of domesticated crea-tures. With creatures like *Drosophila*, which had many forms and which continually produced new ones, it became impossible to devise rational formulas. This was not the first time a system of genetic classification had collapsed when new organisms were brought into play. Mendel's original bipolar system of dominant and recessive forms had been de-signed for an organism, the domesticated pea, in which only two types were known of any trait (e.g., wrinkled and round, yellow and green, short and tall). This system broke down when, in the early 1900s, neo-Mendelians began to apply it to mice and other animals that had char-acters that came in more than two forms (e.g., coat color). The new system of presence and absence formulas was invented to accommo-date this larger number of forms. When Morgan and his students be-gan to use *Drosophila*, with its numerous mutant forms, the presence and absence system failed in its turn.

These shifts from one organism to another—peas to mice, mice to flies—turned out to be unexpectedly subversive of established prac-tices. Drosophilists were the first to encounter the limits of the neo-

Mendelian system because they were the only ones whose breeding experiments were big enough to produce new mutants. Mendelians who worked with mice or fowl had no such experience, because new mutants appeared infrequently if at all in their experiments. With these creatures, apparently definitive formulas could be constructed for each organ series; with *Drosophila* they could not, and the whole system of experimental heredity collapsed and was replaced by a new one. The decisive advantage of the new system of naming and classifying genes by location was its infinite capacity to accommodate as many mutant forms as the breeder reactor could produce.

The organ-series system failed quite unexpectedly in the spring of 1912, just when Morgan was constructing what he thought would be definitive formulas for several organ systems, especially eye color. He was "head over ears in my flies," he wrote Driesch in January:

> They are yielding such a harvest of results that I am swamped. I sent you a paper on eye color a few days ago and now orange has dropped out of the series so that I can carry the results much further. I have a parallel series on body color and on wings. Then the big question of "association" [i.e., linkage] is coming up that will give a splendid chance for even further analytical work.[6]

The two latest eye-color mutants, *orange* and *eosin,* had turned up in February and August 1911 as the fourth and fifth members of the series. Although it meant revising his formulas yet again—it was at least the third time—Morgan was hopeful because he though that eosin resulted from the loss of the orange factor, the existence of which he had predicted but which had so far eluded him.[7] After two years of labor Morgan felt that a definitive formulation was finally within his grasp.

Within months, however, Morgan's hopes for the eye-color series had totally collapsed. The discovery in February and March 1912 of two more eye-color mutants, *purple* and *maroon,* dashed all hope of a definitive formula for that series. As Bridges and Morgan recalled: "Each fresh appearance required that the entire edifice be rebuilt. The system finally became so unwieldy that it had to be abandoned in toto. The present simple system, capable of almost indefinite expansion,

6. Morgan to Driesch, 1 Jan. 1912, THM-APS.

7. C. B. Bridges and T. H. Morgan, *The Third-Chromosome Group of Mutant Characters of Drosophila melanogaster,* CIW publication no. 327 (Washington, 1923), p. 46; and A. H. Sturtevant, "The Himalayan rabbit case, with some considerations on multiple allelomorphs," *Amer. Nat.* 47 (1913): 234–39, on p. 237.

took its place."[8] The organ series failed even more dramatically in experiments—huge experiments—that were designed to produce mutant types predicted by the formulas. Morgan did one such experiment in early 1912 on *pink* and *vermilion*. Enormous numbers of flies were raised, but the predicted new mutant types did not turn up. Morgan was similarly frustrated in experiments with the smaller series of body-color mutants, *black, brown,* and *yellow*. The neo-Mendelian calculus predicted that crosses of black and yellow should produce a new combination of present and absent factors, but crosses involving almost two hundred thousand flies failed to produce the expected new mutant type. Long after the event H. J. Muller recalled his astonishment that Morgan would persist so long in what he, Muller, recognized as a misguided endeavor.[9]

Exactly what was the last straw for Morgan we do not know, and it does not really matter. If it was not the eye- or body-color mutants it would have been something else, and soon. The number of wing-length mutants had grown from six to ten between December 1911 and January 1912, and another body-color mutant, *ebony,* turned up in March. The relentless flood of mutants from the breeder reactor left Morgan no choice but to adopt a fundamentally new system of naming and classifying factors.

On 5 March 1912 Morgan met with Sturtevant and Bridges and gave the green light to their plan for a major campaign of mapping chromosomes 2 and 3 of *D. melanogaster*. This meeting occurred just one day after Bridges established the first case of linkage in an autosome (all previous cases were in the X, or sex, chromosome). Bridges and Sturtevant divided up the known autosomal mutants between them and set to work constructing the stocks that would be required for systematic mapping.[10] The speed of the decision suggests that Mor-

8. Bridges and Morgan, *Third-Chromosome Group*, p. 46 (quotation); and C. B. Bridges and T. H. Morgan, "The Second-Chromosome Group of Mutant Characters," in *Contributions to the Genetics of Drosophila melanogaster,* CIW publication no. 278 (Washington, 1919), pp. 125–342, on p. 169. At about the same time, crosses of *eosin* and *white* were leading to the conclusion—Sturtevant was the first to suggest it—that these two mutations were different forms of one and the same gene, an impossibility in the old system.

9. Bridges and Morgan, "Second-Chromosome Group," p. 148; T. H. Morgan, "Heredity of body color in *Drosophila,*" *J. Exp. Zool.* 13 (1912): 27–44; Bridges and Morgan, *Third-Chromosome Group*, p. 46; and Muller to Edgar Altenburg, 6 Mar. 1946, HJM-A.

10. Bridges and Morgan, "Second-Chromosome Group," p. 171.

gan and his students had planned the mapping project earlier and were just waiting to be sure it was indeed doable.

The decision to proceed with systematic mapping of chromosomes 2 and 3 marks the turning point in the construction of *Drosophila*. Sturtevant and Bridges committed themselves to an instrument and a system of naming and classifying mutants that was good for genetic mapping and for little else. They also committed themselves to a system that produced quantitative knowledge and that would entail major investments in standard tools and a mass-production technology. Within a year or so the mapping project had displaced all rival modes of practice. Locating genes, which had been a minor element in Morgan's neo-Mendelian practice, became the key to the new practice of experimental genetics and to the reconstruction of *Drosophila*.

Constructing *Drosophila*: Boss and Boys.

Mapping did not immediately replace neo-Mendelian work in Morgan's lab. For example, vestiges of neo-Mendelian terminology appear in Sturtevant's first published map of the X chromosome in early 1913.[11] Morgan himself continued to work in the older way for some time after his students had abandoned it. He was not uninterested in the location of genes, just more interested in their developmental relations.[12] It is not clear, in fact, that Morgan expected to participate directly in the mapping project when he and his students planned it on 5 March 1912. We will never know for sure, but it is likely that Morgan regarded the mapping project as the dissertation work of two precocious graduate students, and that he himself planned to focus on what seemed to him the more important task of constructing organ series. How could Morgan forsee that mapping would get so out of hand that it would crowd his own work to the margins?

This generational gap in the early mapping work made a great impression on his young students, who recast it in their memory as a dispute over priority for the key concepts of the new genetics. Muller and Edgar Altenburg were the most insistent that Morgan had received too much credit. He was "badly confused on fundamentals," Altenburg recalled, and Muller had to set him right on multiple gene analysis, interference, and so on. Muller recalled obsessively how Morgan had been slower to understand the new ideas than his young students (especially

11. Sturtevant, "Linear arrangement," pp. 45–46.
12. Morgan, "An attempt to analyze," pp. 384–85, 403–404, 406–407.

himself).[13] It is clear, however, that Morgan and his students simply had different conceptions of what experimental heredity was all about, especially in the critical months when the choice of alternative modes was most unsettled. Morgan certainly understood the new principles of gene mapping and multigene analysis; he was just less willing than his young students to conclude that mapping was more important than the developmental effects of different combinations of related genes.

Morgan had good reasons not to abandon organ series for linkage groups, reasons his students did not fully share or appreciate. Heredity to him was about development and evolution, not genetic maps. Besides, the new system of classification, while it facilitated mapping, seemed dangerously akin to the theoretical conceptions of factor theorists like August Weismann, of whom Morgan was very critical. Identifying mutant types with single genes and giving them the same name (pink eye, pink gene) seemed to resurrect the discredited idea that traits were determined by single factors. The new system of classifying mutants put the whole enterprise of experimental heredity at risk of attack from conservative morphologists, as Morgan knew all too well from personal experience.[14]

Morgan's young students had little of his intellectual baggage and were less alert to the costs and risks of giving up the old system of classification and practice. They thus never hesitated to adopt the narrower, if also more productive, mode of experimental practice. They never missed the biological richness of the older style of heredity, as Morgan did; the choice of competing modes of practice was easy. Morgan hesitated to embrace the new not because he failed to understand it but because he understood all too well what would have to be given up. Asking who deserved credit for what is asking the wrong question.

What we want to know is how students in their early twenties—Bridges and Sturtevant were still college seniors in March 1912—contributed as equals with their far more experienced mentor. That would have been more difficult in most lines of experimental biology, where well-worked and elaborated experimental systems had produced layer upon layer of knowledge. In such fields age and experience were advantages in getting access to the research front. It took a long appren-

13. Altenburg to Muller, 1, 4, 24 Mar. 1946, and Muller to Altenburg, 5, 6 Mar. 1946, all in HJM-A; Carlson, *Muller,* pp. 61–62, 75–76; Allen, *Morgan,* pp. 202–208; and Altenburg to Schultz, 12 June 1961, JS.

14. Morgan, "Factors and unit characters," pp. 9, 14; and T. H. Morgan, "The mechanism of heredity as indicated by the inheritance of linked characters," *Pop. Sci. Mon.* 84 (1914): 5–16, on p. 15.

ticeship to master a vast and arcane literature and to learn the craft
skills to make experiments work. *Drosophila* was different, and the question
is, what specific characteristics of genetic mapping in *Drosophila*
gave the edge to inexperience and lack of sophistication? What enabled
Morgan's "boys" to lead so decisively that Morgan could seem to
them in hindsight to have been more a sheet anchor than a sail? And
how did the youthfulness of those who constructed *Drosophila* shape
the character of the new genetics?

Modern genetics was in many ways an ideal activity for ambitious
and inexperienced newcomers to experimental biology. It was far less
complicated than experimental morphology, and it put a premium on
abstract analytical skills, which are more easily mastered by smart but
inexperienced beginners. It did not require mastery of arcane knowledge
and craft skills. The literature on neo-Mendelian formulas was
virtually irrelevant to mapping, and most of the experimental tradecraft
of mapping would be invented *de novo*. Most important, hunting
and mapping mutants was extremely productive work, ideal for young
people just launching their careers. It played to their advantages,
rewarding quickness and simplicity more than experience and complexity
of vision.

To be sure, Sturtevant, Bridges, and Muller had brains and talent.
However, what allowed them to take the lead in inventing *Drosophila*
genetics was that this mode of experimental practice originated as a
dissertation project for graduate students. The "boys" were in charge
from the start in a project that had to be straightforward and doable
in a finite time because it was a student project. If my interpretation is
correct, the mapping project was meant to complement, not replace,
other modes of experimental heredity. In short, genetic mapping and
chromosomal mechanics were designed by young persons to be a
young person's game. The vast enterprise of modern genetics unexpectedly
took shape around this narrowly defined project. How could
anyone foresee that a project of two college seniors would grow to become
the life work of generations of fly people?

Most of the basic mapping of *Drosophila* from 1912 on was done by
Sturtevant and Bridges, with Muller and others contributing linkage
data and craft tricks as a by-product of their researches on various aspects
of chromosomal mechanics. Morgan also participated, but more
and more he let Bridges and Sturtevant design his genetic experiments
for him.[15] It was Sturtevant and Bridges who invented the basic map-

15. Allen, *Morgan*, pp. 194–95; and Sturtevant to T. Sonneborn, 5 May
1967, AHS 3.19.

ping techniques and rebuilt *Drosophila* into a standard instrument of genetic mapping.

Mapping Technology

Genetic mapping is in principle rather like the triangulation method of topographical mapping. The first step is to establish a baseline by choosing two genes and measuring the distance between them very accurately, by counting large numbers of recombinants. This baseline then serves as a reference to which all other points are related. Mutants are chosen for those base points that are convenient to handle and give crossover ratios close to Mendelian theory: Sturtevant and Bridges selected *black* and *pink* as the base points of chromosomes 2 and 3, respectively. Other marker genes are then selected at positions along the full length of the chromosomes and measured accurately with respect to the base points. Within this framework genes are mapped as they turn up, using whatever markers are nearby on the chromosome. Gene maps are thus built up gradually of interconnecting segments from a baseline, just as a topographical map is built up from a baseline in a network of connected triangles.

Measuring distances along chromosomes is rather less direct than the measurement of distances along the ground, however. In genetic mapping the meaning of "distance" is less concrete and transparent than it is in geodesy. What is actually measured is rates of recombination, or crossing-over, between two genes. Distance is by definition the ratio (expressed as a percentage) of the actual rate of recombination to the rate expected for unlinked genes. The actual work of measuring distance thus consists of separating and counting the progeny of crosses in which recombination has and has not occurred, and taking the ratio of the recombinant classes to the total. This is an idealized concept of distance: it assumes that the rate of crossing-over is constant over the length of a chromosome, when in fact it seldom is constant. It also assumes that the rate of crossing-over depends only on distance, when in fact it is quite variable. Indeed, almost none of the early *Drosophila* stocks acted in this ideal fashion. Also, the raw data must be corrected for double crossing-over, which becomes significant as the distance between two genes increases. The concepts of measurement and distance are artificial constructs, no less than *Drosophila* itself, and nature had to be forced into congruence with theoretical ideals.

Constructing the basic framework of chromosomes 2 and 3 took about two years. Bridges identified the first linkage in 3 (*pink* and *ma-*

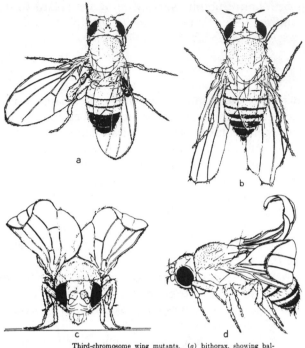

Third-chromosome wing mutants. (a) bithorax, showing balancers turned into extra wings; (b) typical Beaded of stock; (c) curled wing, front view; (d) curled wing, side view.

Figure 3.2   Mutants of the sort used for mapping chromosomes. From Bridges and Morgan, *Third-Chromosome Group*, p. 31.

roon) in April 1912, but the data was suspect, and the first unambiguous linkages were produced in July by Sturtevant, with *pink* and *ebony*, and by Bridges with *pink* and *kidney*. Bridges was able to get quantitative linkage data for *kidney* by November, but Sturtevant was less lucky with his *ebony* stock, which unbeknownst to him contained a hidden gene that suppressed crossing-over. It took him over eight months of hard labor to construct the double-recessive stock that he needed for the linkage measurement. In those early years every new type of genetic noise—lethals, suppressors, modifiers—had to be identified and eliminated, one by one. *Drosophila* had, so to speak, to be debugged.

The map of chromosome 3 began to be useful in December 1912, when the distances between *black, purple, vestigial,* and *curved* were established. *Sepia, spineless, kidney, spread,* and *ebony* were established as the basic framework of chromosome 3 in the course of 1913. By January 1914 the maps of chromosomes 1 (or X), 2 and 3 were fairly well filled in, and they were published in the group's famous four-author book in 1915 (see figure 3.1).[16] These first maps kept changing as mapping stocks and methods were steadily improved, but the changes were generally small and incremental.

We can see in the gradual assembling of these maps the construction of *Drosophila* as a standard experimental instrument. As engineering designs are altered and stabilized as they enter into actual production, so, too, with *Drosophila*. Chromosome maps were the blueprints of a system of mass production that changed and improved through being worked. Experiments were scaled up to produce more precise measures of distance (extremely large numbers of flies are required for closely linked genes), and larger experiments generated lots of new mutants that had to be mapped and assimilated into the mapping process. it was this autocatalytic character of large-scale mapping, plus the requirement of precise quantitative data, that drove the construction of *Drosophila* toward a standard form.

Large-scale mapping also required efficient, standardized procedures, and most of the basic recipes and procedures of modern genetics were invented in the first five or six years of the mapping project. Mass production created an epidemic of technical invention, a mutating period that fed on itself, autocatalytically. The more mutants there were to map, the greater was the pressure to devise more convenient and efficient methods for doing crosses and processing flies. With more efficient methods, drosophilists could scale up their experiments, which produced more mutants, which required still more efficient methods, and so on. The standard maps of 1919–23 represented data from some ten million flies, and altogether about thirteen to twenty million flies were etherized, examined, sorted, and processed![17] How could standardization and efficiency not have become crucial issues in work on such a scale?

The key mapping procedure was the *backcross* (figure 3.3). Say, for example, we want to measure the distance between *black* (*b*) and *vestigial* (*v*). We would first construct a stock of homozygous double reces-

16. Bridges and Morgan, *Third-Chromosome Group*, pp. 48–49, 51–52, 54–55; Bridges and Morgan, "Second-Chromosome Group," pp. 144–45, 156–58, 192–93, 297–303; Morgan to R. S. Woodward, 14 July 1915, CIW; and T. H. Morgan, A. H. Sturtevant, H. J. Muller, and C. B. Bridges, *The Mechanism of Mendelian Heredity* (New York: Holt, 1915), p. ii.

17. T. H. Morgan, C. B. Bridges, and A. H. Sturtevant, "The Genetics of *Drosophila*," *Bibliographia Genetica* 2 (1925): 1–262, on pp. 22, 109.

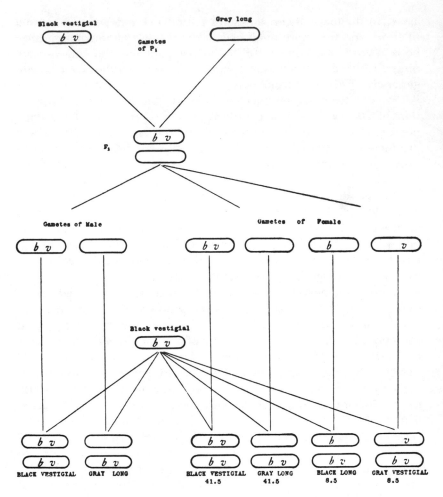

Figure 3.3   The backcross test for measuring the distances between linked genes. From Morgan et al., *Mechanism of Mendelian Heredity*, p. 50.

sive flies (*bv bv*); it is the analytical reagent that will make crossing over visible. We then cross this stock with wild-type flies, symbolized as (++ ++), and collect the female first-generation (F_1) progeny. It is in these heterozygous (*bv* ++) females that crossing over will occur during the process of gametogenesis, to produce eggs that are (*b*+) and (+*v*). The trick is to make these recombinants visible, and that is accomplished by backcrossing the F_1 females with ($bv+$ bv) males. Among the offspring of this backcross (F_2) will be some flies that are either black or vestigial but not both; these flies, genetically ($b+$ bv) and ($+v$ bv), are those in which recombination has occurred.[18] Double-recessive stocks are thus analytical reagents, like the colorimetric reagents of analytical chemists, which make hidden genetic elements visible to the eye.

It was Bridges who transformed the idea of the backcross into a workable and reliable experimental procedure, and he had two lucky breaks. First, a double recessive fell into his lap when *purple* appeared in a stock of *vestigial* and turned out to be conveniently linked to *vestigial* in chromosome 2. Bridges might otherwise have spent many months constructing a double-recessive stock, as Sturtevant was then having to do with *ebony*. Bridges's second lucky break was his discovery that crossing-over does not occur in male drosophilas—a biological quirk, which, had it gone unnoticed, would have made it impossible to get reproducible linkage data. Bridges had happened to begin some tests just as the Morgan group began their annual summer trek to Woods Hole. His flies began the trip as pupae, but en route they hatched and within hours were mating promiscuously. Discarding his ruined females, Bridges continued with his males but found, to his great surprise, no crossing-over at all. From then on only female F_1 flies were used in the backcross procedure. By further standardizing experimental conditions and clearing his reagent stocks of "odds and ends of mutants," Bridges made the backcross into a convenient and reliable mapping tool.[19]

Most standard procedures were devised in the early years of the mapping project; for example, the three-point cross, which was first used with *black, purple,* and *vestigial* in May 1913. Although it was difficult and time-consuming to construct triply marked reagent stocks (it took Bridges almost a year to make the first one), the investment paid off in the long term. Three-point crosses produced data on three points simultaneously and, as a bonus, frequencies of double crossing-over, which were essential for figuring map distances. Also, the data were produced under identical experimental conditions and so could be safely assumed to be internally consistent, which was never the case with data from different experiments.[20] As with any method of large-scale production, it made economic sense to invest up front in tools

18. Morgan et al., *Mechanism of Mendelian Heredity,* pp. 50–51 and fig. 22.

19. Bridges and Morgan, "Second-Chromosome Group," pp. 171–76.

20. Ibid., pp. 185–89.

that would ensure more efficient and precise production in the long run. In 1914 Muller constructed two stocks with five markers, *peple* and *seple,* as a preliminary to his study of the mechanism of crossing-over. With these multiple reagents it was possible to locate genes anywhere on chromosomes 2 and 3, and since the markers were closer together there was no need to correct for double crossing-over.[21] Such reagents became very popular. In 1916 Bridges introduced *Xple,* a reagent for mapping the X chromosome, which was synthesized from six of the best markers then available selected to lie evenly spaced along the chromosome. An improved model of *Xple* came on line in 1920 with a new addition that filled a gap in the original and two substitutions that improved viability and made it easier to separate recombinant classes. Bridges also took the opportunity of a model change to cleanse the stock of modifier genes, so that all regions of the chromosome gave comparable crossover frequencies.[22] Here, indeed, was a standardized reagent for genetic mapping: versatile, quick, dependable, foolproof. Bridges constructed similar multigene reagents, *IIple* and *IIIple,* for mapping chromosomes 2 and 3.

Occasionally mutants turned up that triggered major changes in mapping technology. Just after the first maps were published in 1915, for example, Bridges found two autosomal dominants, *Star* and *Dichaete,* which were ideal markers because their presence in heterozygotes could be determined by inspection, eliminating the second step in the backcross (figure 3.4). That, and their excellent viability, fecundity, and ease of separation from the wild type, made *Star* and *Dichaete* ideal for mapping, and they quickly displaced *black* and *pink* as the anchor points of chromosomes 2 and 3, even though it meant laboriously reconstructing both maps from the ground up. Bridges first synthesized new *IIple* and *IIIple* reagents around *Star* and *Dichaete,* then proceeded to redo all the linkage work on chromosomes 2 and 3, recomputing the maps around *Star* and *Dichaete* as reference points. It was an enormous task, but Bridges anticipated that these improved maps would require little further improvement. But within just a decade Bridges was again forced by the piling up of new data to entirely recalculate the standard maps. It was a chore he did not relish: "I am recalculating the whole damn Chrom II," he complained to Curt Stern.

21. Bridges and Morgan, *Third-Chromosome Group,* pp. 75–77; and H. J. Muller, "The mechanism of crossing over, III," *Amer. Nat.* 50 (1916): 350–66.

22. C. B. Bridges and T. M. Olbrycht, "The multiple stock 'Xple' and its use," *Genetics* 11 (1926): 41–56.

"Basic map comes out around 92 long instead of 107—not finished.[23]

There was a limit beyond which investment in fancy reagent stocks did not pay off in increased production. Muller constructed mapping stocks with up to twenty markers, permitting whole chromosomes to be mapped very accurately in single gigantic experiments.[24] They found little use in everyday practice, however; they were just too complicated to be practicable. Bridges, for whom mapping was his main job, was a shrewder and more pragmatic calculator of the costs and benefits of investment in complex tools. His reagents balanced elegance and practicality, and they were the tools that drosophilists adopted. Muller and Bridges were like the two familiar types of engineer: the one striving for the perfect technology irrespective of cost, the other settling for the technology that was just good enough for cost-effective production. Most of the fly people preferred good-enough technology.

These artfully fashioned mapping stocks exemplify the constructed nature of *Drosophila*. They constituted a new variety of *Drosophila*, one that could exist only in the domestic ecology of the laboratory, serving its human symbionts' desire for quick, convenient, and cost-effective experimental production, and being dispersed to new niches in laboratories around the world. This standard *Drosophila* could only have evolved in the context of a genetic mapping project, because at the time mapping was the only form of experimental breeding that was quantitative. Other, nonquantitative modes, such as neo-Mendelian experimental heredity, gave little incentive to invest so prodigiously in standard organisms and elaborate technologies of mass production.

Controlling Variability

When drosophilas crossed the threshold between the natural and domesticated worlds, they brought with them into the laboratory all the

23. Bridges to Stern, 28 Oct. 1933, CS; Bridges and Morgan, "Second-Chromosome Group," pp. 259–73; Bridges and Morgan, *Third-Chromosome Group*, pp. 127–34; C. B. Bridges, "Current maps of the location of the mutant genes of *D. melanogaster*," *Proc. Nat. Acad. Sci.* 7 (1921): 127–32; Morgan to J. L. Collins, 1 May 1918, UCDG; and Bridges to M. Demerec, 13 Apr. 1933, MD. For the uppercase letters see note 4, above.

24. Carlson, *Muller*, pp. 42–43, 71–82, 86.

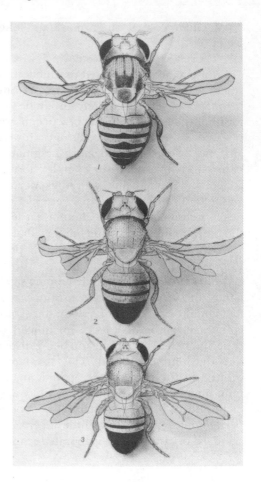

Figure 3.4 *Dichaete* mutants, used by Calvin Bridges to construct improved mapping stocks. From Bridges and Morgan, *Third-Chromosome Group,* p. 166.

natural wildness and variability of wild creatures. Natural selection and genetic drift had tinkered them into a genetic bricolage, harboring every kind of hidden genetic junk—modifiers, suppressors, lethals, and other odd genes—which did little harm in nature but caused drosophilists no end of trouble. Drosophilas also had built into them a great deal of physiological variability. In the wild this ability to adapt rapidly to changing conditions was an advantage, a demographic insurance policy, but in genetic mapping projects it was a real liability. Experiments could produce quite different results from one day to the next, depending on the temperature of the lab, the condition of the fly food, and the age and population density of the cultures. This vari-

ability was not a problem in experiments that were designed to construct formulas for organ groups. It became a very serious problem,
however, in experiments that were designed to produce precise, quantitative data on frequencies of crossing-over. As Morgan observed, every
unexplained anomaly was potentially a stick for the enemies of *Drosophila* genetics to beat him with. "It is surprising," he told a friend, "how
many controls must be made before one can feel secure that the location of a factor . . . is correct, or fairly so."[25]

One could to some extent simply avoid stocks that were troublesome, and the rapid accumulation of new mutants offered mappers an
ever-wider selection from which to choose. Bridges remarked on the
"continual improvement of the working material by the substitution of
better mutants."[26] Good mapping mutants were well defined and easy
to separate quickly and unambiguously from wild type and related mutants; they were vigorous, ready to mate, and fecund of viable offspring.
They were also "free from such bad habits as getting drowned, or stuck
in the food, or refusing to be emptied from the culture bottle, etc.,
which alienate the affections of the experimenter."[27]

Some variability could not be avoided, however, because some of
the best mapping mutants were also the most variable and difficult to
use. Mutants that were extreme were easiest to separate from the wild
type, yet they also tended to be less vigorous and to smother in their
food or otherwise die before they could be counted. Uniting several mutant genes in a single stock made it more useful but also less
viable. Thus constructing good mapping stocks became a balancing
act, in which benefits in efficiency and convenience were pur-
chased at the cost of diminished vigor. Controlling variability was perhaps the most important element in the construction of a standard
Drosophila.

It is useful to distinguish three different ways in which drosophilists
tried to control variability: literary, environmental, and genetic. Gene
mappers could simply accept variability as an unavoidable fact of experimental practice and try to compensate for it cosmetically by disclosing
their methods or by applying correction factors to data they knew were
imperfect. Alternatively, they could accept variability as a biological fact
and do what they could to reduce it by standardizing the conditions in
which experiments were done in the lab. And finally, mappers could

25. Morgan to Collins, 1 May 1918, UCDG; and Morgan, "Program of
work," 2 Oct. 1917, CIW.

26. Bridges, "Gametic and observed ratios," pp. 57–58, 61.

27. Bridges and Morgan, "Second-Chromosome Group," p. 134.

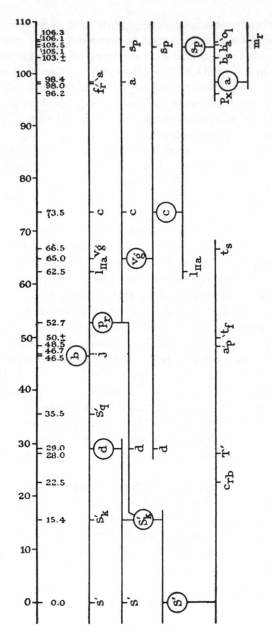

Figure 3.5 "Constructional" map, showing the sequence in which map segments were measured and assembled. From Bridges and Morgan, "Second-Chromosome Group," p. 302.

reconstruct *Drosophila* genetically to get rid of the noise—the ultimate solution to the problem of variability.

A good example of the strategy of literary disclosure is the so-called constructional map, in which the traditional linear map was expanded in a second dimension to reveal the order in which segments of map had been assembled (figure 3.5). Since map positions were only as accurate as the base points from which they were measured, knowing how a map was constructed would help users judge the trustworthiness of linkage data. (Confidence could also be inspired by indicating the more reliable data by heavier lines.)[28] A historical map was no remedy for variability, but it was one way of maintaining trust in the enterprise at a time when production methods were improving rapidly and when early errors were likely to be publicly exposed. This somewhat awkward representation never caught on, but something like it did come into common use as a private research tool. This was the "working and valuation map," in which map location of mutants was indicated on the vertical axis and their value as research tools (ease of use, reliability, certainty of location) by position on a horizontal axis (figure 3.6). Mounted on fiberboard with labeled thumbtacks indicating genes, these working maps could easily be updated as the positions and usefulness of mutants changed. Bridges had a large four-sided working map, one side for each chromosome of *D. melanogaster,* hanging in easy reach just behind his desk.[29]

Another way of disclosing variability without eliminating it was to apply correction factors to data that were acknowledged to be imperfect. Bridges and Sturtevant tried for a time to calculate coefficients of viability for each one of their mapping stocks, rather as physical chemists applied activity coefficients to get real concentrations from measurements of unideal solutions.[30] It was a makeshift, and not surprisingly it never caught on. Standard stocks changed too fast, and viability was as much a property of experimental conditions as of genetic constitution.

A strategy of literary disclosure may also have inspired the peculiar narrative form of the three monographs that Bridges and Morgan published between 1916 and 1923, in which all their results were presented in chronological order, replicating the history of how mutants

28. Ibid., pp. 302–303.

29. Ibid.

30. C. B. Bridges and A. H. Sturtevant, "A new gene in the second chromosome of *Drosophila* and some considerations on differential viability," *Biol. Bull.* 26 (1914): 205–12.

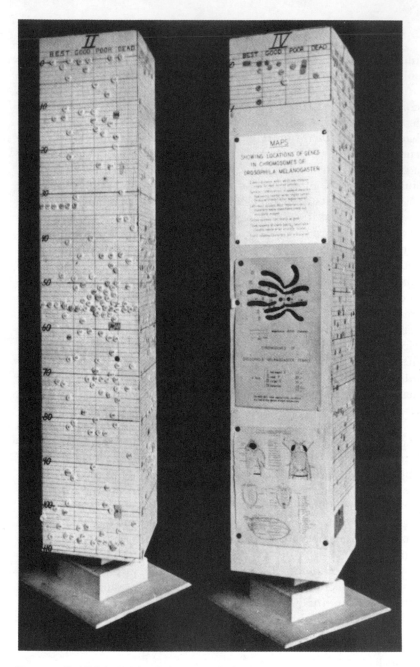

Figure 3.6   Calvin Bridges's "totem pole," a four-sided working and valuation map showing locations of mutant genes and their relative usefulness for mapping. From T. H. Morgan, "Personal recollections of Calvin B. Bridges," *Journal of Heredity* 30 (1939): 355–58, on p. 356.

and maps had accreted. It was and still is extremely rare for scientists to present knowledge in this form. As a rule, history is banned from scientific reports, and a prolix style has not been used to give readers a sense of "virtual witnessing" of experiments since the days of Robert Boyle three hundred years ago.[31] Doubtless there were many reasons why Morgan made results public in this unusual way (it was the simplest way of presenting very large amounts of data that few people would ever use). However, the historical narrative and prolix detail of these monographs may have been meant, *inter alia,* to remind readers that if they did not get the same results it was because the technology had evolved, not because Morgan and company were sloppy and untrust-worthy workers.

The second, or environmental strategy—controlling the environ-mental conditions of mapping experiments—was the most cost-effective way of reducing variability, especially in the early years of the mapping project. One early and very important improvement was to breed only one pair of flies in each bottle. In the older style of mass culture, starvation and severe competition caused by overcrowding re-sulted in poor viability and highly variable results. Cultivation under optimal conditions of food, temperature, and population density made crossing-over experiments much more reproducible. Such improve-ments meant more work, more culture bottles to handle and clean, more food to prepare, more precautions in checking incubators. How-ever, the extra investment was justified by the better quality of the data produced.[32] Another way of finessing variability was to redesign cross-ing experiments so that the effects of low viability would statistic-ally cancel out. In the technique of "complementary crosses," for ex-ample, two sets of F_1 flies were constructed in such a way that a gene causing poor viability would end up in opposite classes in the back-crossed F_2. By pooling the crossover and noncrossover classes from the two backcrosses, errors due to inviability would cancel out in the ratio. Variation in crossing-over was also reduced by counting off-spring only from the first ten days of egg laying. Bridges had hoped to increase the productivity of crosses by counting the offspring over twenty or even forty days rather than the usual ten or so, but found

31. Thomas S. Kuhn, *The Structure of Scientific Revolutions* (Chicago: Univer-sity of Chicago Press, 1962), ch. 11; and Steven Shapin, "Pump and circum-stance: Robert Boyle's literary technology," *Soc. Stud. Sci.* 14 (1984): 481–520, on pp. 491–93.

32. C. B. Bridges, "Gametic and observed ratios in *Drosophila,*" *Amer. Nat.* 55 (1921): 51–61; and Bridges and Morgan, *Third-Chromosome Group,* pp. 70–71.

that the rate of crossing-over decreased in the offspring of older mothers.[33]

Environmental controls constrained the expression of *Drosophila*'s natural variability, but they did not eliminate its biological causes. To do that, drosophilists had to rebuild the creature's genetic machinery—the third, genetic strategy. Hidden lethals were recognized when they resulted in the absence of entire classes of recombinants; they could then be gotten rid of by outcrossing with wild stocks and selecting "normal" offspring, that is, those that displayed Mendelian ratios. Genes that modified or suppressed crossing-over in nearby loci were harder to detect and could be eliminated only by rebuilding the chromosomes. That was done by selecting a dozen or so stocks in which the rate of crossing-over was least variable and selectively breeding from them a standard stock in which the rate of crossing-over was uniform over a given stretch of chromosome.[34] In other words, usable bits were extracted from a dozen stocks and reassembled in a single genetic bricolage. Wild drosophilas were thus physically reconstructed to conform with the fundamental principles of genetic mapping: namely, that "distance" is proportional to the rate of crossing-over and that the rate of crossing-over is constant from one region of a chromosome to another. It was not so in nature, but it was made so in the lab by constructing "standard" flies.

The reconstruction process was incremental. Take the case of *pink*, for example. Most of the early crosses with *pink* stocks gave anomalous data. Changing from mass to pair cultures improved matters, but one quarter of the crosses still departed from the theoretical. Painstaking study of the conditions of culturing eliminated most of the remaining variability, but still five stocks were anomalous, owing, it turned out, to the presence of a lethal gene that was close to the *pink* locus and thus hard to detect and expunge.[35] In this way was *Drosophila* gradually rebuilt and made to conform to the ideal principles of Mendelian theory and chromosomal cartography. Drosophilists built the purposes of precise, quantitative mapping into the material fabric of their living instru-

33. Bridges and Morgan, "Second-Chromosome Group," pp. 181–83; see also C. B. Bridges, "A linkage variation in *Drosophila*," *J. Exp. Zool.* 19 (1915): 1–21.

34. Bridges and Olbrycht, "Xple," p. 49.

35. Bridges and Morgan, *Third-Chromosome Group*, pp. 46–47, 107–108. For another good example, *purple*, see Bridges and Morgan, "Second-Chromosome Group," pp. 169–99.

ment. What more concrete evidence could we desire of Drosophila's constructed, technological nature?

Fly and Fly People

The construction of *Drosophila* also involved fundamental changes in the relationship between the fly and its human symbionts. It changed experimenters' perceptions of phenomena. Consider, for example, the "epidemics" of mutation that swept the Morgan lab in the early years of the mapping project, when the discovery of a new mutant seemed to trigger the appearance of others like it. These epidemics were in part the result of rapid changes in perceptions and categories of meaning. Finding a new mutant opened fly breeders' eyes to the possible existence of others like it, and since finding new mutants was a mark of achievement, the finding of one mutant was a powerful incentive to find more. Creating a new category invited mappers to perceive as mutants appearances that before they would have passed over as normal and not very interesting variations in the wild type. Once the mappers created a new class of *jaunty* wing, for example, "suddenly in certain other stocks wings that have been passed over as 'imperfectly unfolded' or only vaguely recognized as 'queer' are seen to be sharply characterized 'jaunties.'"[36] In fact, some of these new "mutants" turned out to be just normal variations after all, so eager were mutant hunters to exercise their new skill in observation and to fill their new categories with real objects.

These cases of mistaken identity offer glimpses into the perceptual changes that accompanied the transformation of a wild creature into a domesticated, artificial one. When dealing with more or less wild flies, mappers perceived variations in ways that were appropriate to variable, wild creatures. With domesticated, standard flies, however, mappers perceived not variations but mutations, important objects to preserve and feed into the mapping process. Mistaking mutants for variations, or variations for mutants, was characteristic of a passage in the construction process when the fly was no longer entirely wild but not yet fully standardized.

Mappers' emotional responses to *Drosophila* also changed in the formative years of the mapping project. They referred to their creature

36. Bridges and Morgan, "Second-Chromosome Group," pp. 161–62, 178–79.

in more respectful terms: we cease to hear complaints about "pesky" flies or murmurs of their low habits. Sturtevant, and no doubt others, began to speak familiarly of Drosophila as a "brute" or "beast," and J. B. S. Haldane referred to it as "that noble animal."[37] It is risky to push too hard on such slight evidence, but such terms do suggest real admiration and respect, of the kind we feel for valuable and useful domesticated beasts or for creatures that remain impressively untamed and endowed with virtues we wish we had ourselves. There is little question that domestication elevated the status of *Drosophila* in the hierarchy of value through which humans perceive the animal world. That should not surprise us: So long as the fly was a common found object, there was no reason to put a high value on it. But spend several years laboriously constructing standard mapping stocks, and your attitude is bound to be more respectful. Domesticated *Drosophila* represented a vast investment of time and skill; it was the basis of productive careers and livelihoods, a ticket to high academic status. It was an admirable and noble beast, indeed, that could achieve such good things.

Similar changes in attitude are evident in the names that mappers began to apply to themselves—"the fly people," "Drosophilites," "Drosophilitics," or (most commonly) "Drosophilists."[38] The evidence is fragmentary, but what there is suggests that the drosophilists identified with their creature, as people who live close to nature identify with creatures who seem to embody admirable or protective virtues, or as fanciers and pet owners identify with their favorite creatures.[39] *Drosophila* became, for the Morgan group and others, a kind of totem animal. And well it might, considering the extraordinary benefits that it conferred upon them. For Morgan's young drosophilists it meant scientific fame and fortune at an extraordinarily young age—no wonder they identified personally with the creature that brought them such rewards.

37. Sturtevant to Demerec, 27 Aug. 1927, MD; Sturtevant to O. Mohr, 2 Dec. 1919, 4 Mar. 1921, 3 Dec. 1922, all in OM; and J. B. S. Haldane to L. C. Dunn, 19 Oct. 1932, LCD.

38. Morgan to C. Zeleny, 11 Aug. 1915, CZ; Sturtevant to Stern, 8 Jan. 1927, CS; Sturtevant to Mohr, 16 Feb. 1919, 15 Sept. 1920, OM; and E. A. Anderson to Demerec, 9 Feb. 1924, 22 Mar. 1927, MD.

39. James Serpell, *In the Company of Animals* (Oxford: Basil Blackwell, 1986).

Stock Keeping and Maintenance

The central feature of a *Drosophila* lab was a good collection of mutant stocks—it was the drosophilists' indispensable tool, their bread and butter. Such collections were the repositories of mutants, which if lost might never reappear, and of constructed reagent stocks, which were not irreplaceable but troublesome to reconstitute should they die out.[40] Maintaining a comprehensive stock collection was no small task. By 1914 the Morgan group had over 100 mutants in cultivation, and by 1925 the number had grown to 395.

The drosophilists' stock collection was not a museum of type specimens but a dynamic, ever-mutating instrument that reflected the ever-changing evolution of the research front. The inhabitants of a *Drosophila* stockroom constituted a kind of village society, in which there was a continual rise and fall in reputation and prestige. For example, *kidney* was highly valued in the early mapping work but was displaced by *spineless* and *ebony*, its near neighbors on chromosome 3, because the kidney eye shape became less extreme as cultures aged and thus harder to distinguish from wild type. (It would have been thrown out had it not been so inextricably part of standard reagents like *peple* and *seple* and their descendants.)[41] Similarly, *sepia* was displaced from its preeminent position at the far left end of chromosome 3 by *Hairy*, which was dominant and did not modify other eye colors. *Pink* was replaced by *peach*, which was more easily distinguished from wild type, and *peach* was in turn replaced by *apricot*. It was shirtsleeves to shirtsleeves in thirty generations. Superior merit was not always sufficient to beat out inherited position in *Drosophila*'s village society, however. *Band*, though a good mutant, had the misfortune to be too close to *ebony* and *Hairless*.[42] *Jaunty* was as good a producer as *black*, but *black* continued to be used "because of its greater prestige [and] because it already exists in many useful combinations."[43] Some inconspicuous mutants found favor despite the difficulty of separating them from wild type because they were hardy or happened to occupy a blank space on the map.[44] Fashions in

40. Demerec, "Maintenance of Drosophila melanogaster stocks," to W. Gilbert, 25 Apr. 1941; and Demerec to V. Bush, 25 Nov. 1940; both in CIW file Drosophila Stock Centers.

41. Bridges and Morgan, *Third-Chromosome Group*, p. 78.

42. Ibid., pp. 89, 49–50, 82–85.

43. Bridges and Morgan, "Second-Chromosome Group," pp. 163–64.

44. C. B. Bridges, "White-ocelli: An example of a 'slight' mutant character with normal viability," *Biol. Bull.* 38 (1920): 231–36.

drosophilas sometimes imitated the human variety. A mutant with short bristles was named *bobbed* after Phoebe Reed's newly bobbed hair (it appeared only in females).[45] Keeping up with these continual changes could be a problem. For example, the stocks that Otto Mohr brought back to Oslo in 1919 were state of the art then, but by 1926 were "rather antiquated" and had to be replaced by stocks borrowed from Curt Stern, a more recent visitor to the fly group.[46]

A large stock collection had to be an integral part of an active research program. Keeping stocks was not work that could safely be left to assistants. Technicians could do the routine transfers, but only active researchers had the know-how and incentive to maintain a collection properly. Bridges was responsible for the stock collection in the Morgan group, assisted from time to time by a full-time technician. When Milislav Demerec offered George Beadle a job at Cold Spring Harbor, it was understood that he would help maintain the stocks.[47] It required constant vigilance to keep stocks plentiful and fit for experimental use. Cultures had to be transferred to fresh food every two weeks or so and had to be examined frequently for contamination by unwanted mutations or wild interlopers. Stocks that were not in active use for research tended to deteriorate. Demerec's fine collection of *D. willistoni* stocks gradually disappeared over a period of seven years and were finally wiped out by an epidemic of fly mites. The Morgan group's collection of *D. melanogaster* deteriorated after Bridges died in 1938 because no one was using it for his own research.[48] Maintaining the machinery of mass production was an essential part of systematic genetic mapping.

There were many hazards in stockkeeping, and even experienced drosophilists could get caught out. Incubators could fail, as Ernest Anderson's did just after he left for a Christmas vacation. "The flies," he quipped, "probably thought that winter had begun with a vengeance."[49] Or coolers could spring a leak, releasing toxic sulfur dioxide into the culture. Clarence Oliver was lucky when it happened to him: all his adult flies died, but some larvae survived that had burrowed into the banana mash. Dropping all his research for a rescue operation, Oliver managed to save more than half his cultures and almost all of

45. Sturtevant to Mohr, 29 Jan. 1921, OM.

46. Mohr to Stern, 7 Dec. 1926, CS.

47. Demerec to G. W. Beadle, 28 Dec. 1931, MD.

48. Ibid.; Demerec to Muller, 7 Feb. 1931, MD; Metz to Demerec, 8 July 1929, MD; and Sturtevant to J. R. Page, 13 Sept. 1945, AHS 3.10.

49. E. G. Anderson to Demerec, 30 Dec. 1923, MD.

the ones that were irreplaceable.[50] Mouse attacks were another frequent hazard, sometimes catastrophic. Mice got into E. N. Wentworth's cultures and "either devoured or liberated nearly everything," leaving him with a single mating pair; resourcefully, he decided to work on inbreeding. Muller panicked when mice attacked his stocks, sending everyone in the lab rushing out for mousetraps and poison.[51] Left in charge of Charles Zeleny's stocks for a summer, David Thompson had to cope with mouse attacks, spoiled fly food, and stocks that crashed unpredictably and needed emergency rescues. Amos Hersh had similar experiences: the sight of uninvited wild drosophilas hovering around his supply of bananas made him so worried about contaminants that he at once put every stock under the microscope for examination. Hersh also had to cope with fans stopping in cooling boxes (a real disaster since Zeleny was working on temperature effects).[52]

Even worse than marauding mice were the infestations of fly mites, which sometimes wiped out whole stock collections. Such epidemics seem to have gotten worse as the *Drosophila* network expanded, just as human epidemics thrive on the expansion of trade networks, when locally endemic parasites encounter virgin populations. A remarkable amount of ingenuity was spent on devising methods to prevent such epidemics, including setting culture racks in kerosene to prevent mites from trekking in, and putting immigrant stocks in quarantine before assimilating them into local populations.[53] Friends and visitors became suspect carriers of epidemics. "I pray that you bring your cultures absolutely mite-less," Dobzhansky enjoined Curt Stern. "Jack Schultz of course did bring mites, and [Boris] Spassky and I are trying to delouse him by keeping his cultures temporarily in [quarantine]. . . . I am sure you can delouse yourself just as easily, if any such undignified operation is necessary."[54]

50. C. Oliver to Demerec, 4 Feb. 1938, MD; and Oliver to Muller, 28 Aug. 1938, HJM-C.

51. E. N. Wentworth, "The segregation of fecundity factors in *Drosophila*," *J. Genet.* 3 (1913): 113–20; and P. Koller to Demerec, 8 Mar. 1939, MD.

52. D. Thompson to Zeleny, 7 Sept. 1919, 29 June, 12 July 1920, CZ 5; and A. Hersh to Zeleny, 21 June, 21 July, 16 Aug., 3 Sept. 1921, CZ 17.

53. Dobzhansky to Demerec, 18 Nov. 1936, MD; Metz to Demerec, 18 May 1933, MD; and *DIS* 6 (1936): 67–72.

54. Dobzhansky to Stern, 29 June 1944, CS.

Paraphernalia and Procedures

Large-scale genetic mapping also required equipment that was efficient, convenient, and cheap. Initially, Morgan's equipment was primitive—it was for him almost a point of honor to make do and improvise.
Flies were cultured in a diverse collection of glass containers, anything
that was cheap and available. Half-pint milk bottles were especially well
suited to fly culture, and members of the fly group foraged them from
front stoops on the way to work. Transferring flies was laborious before
ether was introduced as an anesthetic. Fernandus Payne recalled that
Morgan would isolate individual flies by getting half of a culture into
a fresh bottle and repeating the process until the desired fly was
captured, unless of course it had escaped in the process. (Drosophilas, being phototropic, can be coaxed into a bottle by aiming it
at a source of light.)[55] Sturtevant and Bridges did not share Morgan's
austere tastes in equipment, and the ever-larger scale of the mapping project was a powerful incentive to improve the technology
of fly culture. An efficient technology of mass production was
no less a part of the construction of *Drosophila* than the standard fly
itself.

Bridges, as keeper of the stocks, was the one most interested in
developing a technology of mass production, and many—perhaps
most—standard fly techniques and paraphernalia were his inventions.
He spent a great deal of time, for example, trying to improve upon the
traditional culture medium of mashed and partly fermented bananas.
A cheap, standard medium was essential for getting reproduceable results at low cost. Large-scale mapping made bananas a major item of
expense in a fly lab—$100 to $200 a year by 1915.[56] As with any mass-
production technology, there was a tension between improving tried-
and-true methods incrementally and gambling on entirely new and
better ones. Incremental improvement disrupted production less in
the short term, but new methods led to much higher productivity in
the long run. Bridges shifted from one strategy to the other. A completely defined synthetic medium would have been ideal, like those

55. A. H. Sturtevant, "Thomas Hunt Morgan," *Biog. Mem. Nat. Acad. Sci.* 33
(1959): 283–325, on p. 298; C. B. Bridges, "Apparatus and methods for *Drosophila culture*," *Amer. Nat.* 66 (1932): 250–73, on pp. 265–67; Carlson, *Muller,*
pp. 49–50, 75–76; and Morgan et al., *Mechanism of Mendelian Heredity,* pp.
230–31.

56. Allen, *Morgan,* p. 250.

Figure 3.7   Fly culture in
specially designed bottle with
yeasty banana pulp and absor-
bent paper. Courtesy of American Philosophical Society
Library, Stern Papers, photographs file Lab and Flies.

used to culture bacteria, but such a medium proved surprisingly difficult to concoct. Bridges made a concerted effort in 1916–17, probably
in connection with his reconstruction of the maps of chromosomes 2
and 3. He first tried a wholly synthetic medium of salts and sugar, but
drosophilas refused to live on such a Spartan diet. He then tinkered
with the traditional banana medium by adding antiseptics to suppress
molds, but the improvement was minimal, and he returned to the synthetic diet, now fortified by chemical nutrients. Unfortunately, molds
grew faster on that than flies did, so Bridges tried harder to improve
the old banana medium, adapting a banana-juice agar that had been
developed at the Rockefeller Institute for small-scale fly breeding. By
fortifying the banana agar with banana pulp and seeding the surface
lightly with yeast suspension, Bridges finally succeeded in creating a
cheap and precisely defined medium. Bridges switched to the new medium in the spring of 1917, and the rest of the fly group, seeing that it
did not seriously disrupt production, adopted it a few months later.
Banana agar remained the standard medium until the early 1930s,
when Bridges's impulse to tinker and improve again mastered him, and

he finally succeeded in devising an agar medium without banana, using cornmeal and molasses.[57]

Bridges also took the lead in building constant-temperature equipment, which became a necessity as mapping experiments became more precisely quantitative. Cold nights do not kill drosophilas, but greatly slow their rate of egg laying and larval development. Heat was a more lethal threat to *Drosophila* colonies in the days before air conditioning, and care had to be taken on hot summer days, when whole collections could be lost to heat stroke. Bridges's earliest temperature-controlled cabinets (1913) were improvised out of bookcases, incandescent lights, and primitive thermostats. In 1922 he designed improved units that controlled temperature, ventilation, and humidity. These homemade contraptions were cheaper than commercial apparatus, though probably also more temperamental. In the early 1930s Bridges designed a third, improved generation of incubators.[58]

An etherizer for anesthetizing flies was another indispensable item of fly paraphernalia, used every time flies were transferred for inspection and sorting. The procedure had to be done just right: too little ether and the flies would awaken before they were sorted, stagger off, and escape; too much ether and they would never wake up. Bridges's design of a special apparatus for etherizing and transferring flies (1919) was the first in a long series of variants.[59] The variety of ingenious designs of this mundane little gadget reveals just how important it was for large-scale breeding. Bridges also introduced the use of binocular microscopes and special illumination for examining and sorting flies. Though more troublesome to use (Morgan preferred his hand lens), microscopes paid off in more precise separation of mutants, especially those that were hard to distinguish from wild type. The technology of *Drosophila* production and the drosophilists' standard recipes were in continual flux, like the stock collection and its inhabitants. Pe-

57. Bridges, "Gametic and observed ratios," pp. 52–54; C. B. Bridges and H. H. Darby, "Culture media for *Drosophila* and the pH of media," *Amer. Nat.* 67 (1933): 437–72; and T. H. Morgan, "Calvin Blackman Bridges," *Biog. Mem. Nat. Acad. Sci.,* 22 (1941): 31–48, on pp. 34, 39. Molasses and cornmeal were nonstandard commercial products, and not every local variety worked the same; see Otto Mohr to Stern, 5 Oct. 1928, 5 Oct. 1931, CS.

58. Bridges, "Apparatus and methods," pp. 250–56; C. B. Bridges and H. H. Darby, "A system of temperature control," *J. Franklin Inst.* 215 (1933): 723–30; and Bridges to Stern, 28 Oct. 1933, CS.

59. Bridges, "Apparatus and methods for *Drosophila* culture," pp. 268–73; Muller, "Etherizing bottle," *DIS* 2 (1934): 62–63; and *DIS* 6 (1936): 52–56.

riods of gradual improvement were punctuated by episodes of more rapid change, when Bridges undertook to update or rebuild the standard maps or when a major new technique was invented, like X-ray mutation or salivary chromosome cytology. About the only piece of *Drosophila* technology that did not change was the culture bottle itself. Bridges, Muller, Stern and others devised a variety of imaginatively shaped bottles, designed to be packed more closely or to keep food from spilling out; none, however, was better than the cheap, good old-fashioned half-pint milk bottle.[60]

A state-of-the-art *Drosophila* laboratory circa 1940 required a few binocular microscopes, several thousand bottles and vials, a calculating machine, miscellaneous equipment (etherizers, counting plates, fly morgues), one or two incubators, an X-ray setup for making mutants, a constant-temperature room, and a kitchen with autoclave. Several technical assistants might be engaged: a food maker and bottle washer, a stockroom caretaker, and a "virgin catcher," who cleared cultures of new hatchlings before they became sexually active, every eight hours, around the clock, every day. The unskilled jobs were customarily done by student helpers, while the more skilled jobs were assigned to technicians with college degrees in biology or divided up among the research staff.[61] Such a lab was less expensive and complicated to operate than a guinea pig colony or plant-breeding station, but it did require some ability to raise funds and manage people as well as flies.

Conclusion

The construction of *Drosophila,* I have tried to show, was like the construction of any laboratory technology. It involved a choice of alternative product designs—organ formula or chromosome map—and the invention of standard tools and a complex system of mass production to make them work. Wild drosophilas had to be rebuilt to serve drosophilists' purposes. The natural variability of wild flies had to be controlled, first by means of literary devices, which disclosed the practical limitations of mapping with half-wild stocks, then in more concrete and

60. Bridges, "Apparatus and methods for *Drosophila* culture," pp. 265–67; Morgan, "Bridges," pp. 34, 39; Carlson, *Muller,* p. 79; and Dobzhansky to Stern, 6 May 1933, CS.

61. J. Patterson to Dobzhansky, 15 May 1936, TD; Dobzhansky to F. Schrader, 16 Nov. 1939, TD; Muller to Payne, 23 Apr. 1945, HJM-A; and Stern to Dunn, 19 Feb. 1934, CS.

Figure 3.8   A well-appointed *Drosophila* laboratory, vintage mid-1930s. Courtesy of American Philosophical Society Library, Stern Papers, photographs file Lab and Flies.

permanent ways. Standard experimental conditions and procedures were established, aimed at minimizing opportunities for half-wild drosophilas to display their inbred talent of adapting to a changeable environment. Finally, the variability of wild flies was eliminated by constructing artificial varieties that behaved genetically only in ways that conformed to the artificial principles of Mendelian genetics and chromosomal cartography. "Standard" drosophilas were, like chemical reagents or physical instruments, constructed artifacts of laboratory life.

   The construction of *Drosophila* also entailed a substantial investment in machinery for large-scale production: standard mapping stocks, recipes and procedures, paraphernalia, quality controls, and all the thousand and one little things that made *Drosophila* so impressively productive and so different from its wild relations and other laboratory creatures. These practical aspects of experimental work may seem mundane and trivial in comparison with big concepts like the gene, but they were no less essential to realizing *Drosophila*'s potential as an experimental system. The concepts of modern genetics were, in a way, the software of an intricately engineered material culture of production.

The construction of a standard fly was not a necessary or inevitable consequence of *Drosophila*'s intrusion into a laboratory of experimental biology. Rather, I have argued, it was specific to genetic mapping, the one mode of experimental breeding, of all available modes, that was both large scale and quantitative. Mapping shared with the pure-line and neo-Mendelian modes of experiment the self-propelling quality of producing more material for experiment than it consumed. But only mapping required a standardized instrument to produce results that others could reproduce, trust, and use. In mapping alone was the imperative to build a standard organism inherent in the work process. Genetic mapping was invented as a practical system of classification and data management when the neo-Mendelian system lost credibility and failed. The construction of the standard fly was a necessary, if unforeseen, consequence of that practical choice.

The creation of a standard fly by the Morgan group is not the whole story, of course. Standard practices are by definition those that are accepted by communities of practitioners, and we will see in a later chapter how the instruments and procedures of the Morgan group were dispersed to biologists everywhere, becoming truly standard. But it is worth just noting here that it was the program and the technology of mass production that enabled *Drosophila* to disperse so successfully. Systematic genetic mapping generated important problems in chromosomal mechanics with the same fecundity that it generated new mutants, and this fecundity is what transformed a localized mode of practice into a cosmopolitan one. Drosophilists followed the fly, and together they colonized the world.

The construction of *Drosophila* profoundly changed the ecology of laboratories of experimental heredity. *Drosophila* drove other creatures out of these workspaces, creatures like *Phylloxera* that could not compete with the fly's productivity and expansive power. At the same time, drosophilists cut themselves off from biologically richer varieties of experimental heredity, of the sort that Morgan and his students left behind in 1912–14. Linkage maps gave little encouragement to those who might still want to explore the genetics of development. The concepts and visual representations of mapping made it more difficult than before to envision relations between genes and morphological traits.

Drosophila and its competitors were not the only creatures whose way of life was changed; so, too, were the drosophilists themselves. The Columbia fly group evolved a distinctive communal culture, which was shaped by the mapping project and by the peculiar imperatives of doing experiments collectively with such a fecund source of rich material and problems. The fly people developed habits and moral principles that ensured access to their communal instrument and that made it possible for others to help exploit the potential of the breeder reactor—as we will see in the next chapter. None of these consequences were foreseen or intended. Far-reaching changes in the material culture and modes of experimental life followed from apparently innocent changes in experimental organisms and practices.

FOUR

The Fly People

T HE FLY LAB WAS AS DISTINCTIVE a domestic ecology for its human inhabitants, the fly people, as for its dipteran denizens. There drosophilas were reconstructed as "standard" flies, and there, too, students and visitors were transformed into drosophilists and assimilated into a working community with its peculiar customs, ways of life, and moral economy. The fly group's rules of communal behavior constituted the design of an intricate piece of social technology, much as linkage maps constituted the design of a laboratory instrument. *Drosophila* and the drosophilists evolved symbiotically, and the customs of the fly people were adapted to the special qualities of their breeder reactor. *Drosophila* was designed to take advantage of its potential for abundant, fast-paced production, and so, too, was the fly group. The fly group's rules of ownership, access, and credit were shaped by the peculiarities of the creature that they had adopted—or that had adopted them. Fly and fly people became dependent on each other for their identities and livelihoods, bound together in doing experiments.

Social relations among the fly people were also adapted to the conditions of their work together. The fly group was a hybrid institution: a small research institute, with a permanent core of full-time workers, encysted in a large academic department. As in a research institute, the drosophilists were systematically organized for production. They worked on a single organism, not on different ones as was customary in academic departments of biology, and their work was programmatic, revolving around a single large instrument and project: namely the *Drosophila* stockroom and the *Drosophila* mapping project. Departments, in contrast, expanded by diversifying lines of work; long-term projects were neither politically nor financially viable. The drosophilists also worked collectively; everyone meddled in everyone else's work all the time, swapping mutants, ideas, and craft lore. Members of the group were actively discouraged from staking out special lines of work as their personal domains. There was no formal division of labor in the fly group, no dividing up of *Drosophila* genetics into subfields, as was common in academic institutions. Credit for achievement went to those who made experiments work and got papers published, not to those who had bright ideas first. This custom was hard on some, but it maintained moral order in a group that did not have the option of avoiding conflict by working on different organisms or in different fields. Conflict was managed not by avoidance and dispersal of effort but by a moral ethos of cooperation and communality.

Morgan's chief drosophilists were full-time professional employees of the Carnegie Institution, but they remained in some ways Morgan's students, his "boys." They had no academic appointments, though they trained graduate students and postdoctoral fellows. Morgan's boys enjoyed opportunities for full-time production that few academics could hope for, but their habit of deference to "the Boss" made issues of authority and control more complex and ambiguous than was ordinarily the case in either academic department or a research institute. Intergenerational tensions also resulted from the way Morgan recruited his core group exclusively from his own students. Ernest Anderson and Theodosius Dobzhansky, who joined the permanent group in 1928 and 1934, were the first who had not been raised there. Internal recruitment encouraged neoteny in social relations, extending into mid-career the unequal relation between mentor and disciple. While Morgan's "boys" enjoyed virtually complete control of experimental work, they also preserved habits of deference to the Boss that eventually became corrosive of group solidarity. Lingering generational tension was part of the price for privileged access to a marvelous machinery of production.

The Drosophilists

Who were the drosophilists, and how did they acquire their special identity in the world of academic zoology? Most of those whose names we remember got their Ph.D.'s in a few brief years between 1914 and 1917: Sturtevant, Bridges, Muller, Charles Metz, Alexander Weinstein, and Edgar Altenburg (who got his degree in botany but was close to Muller.) A core group worked at Columbia. Sturtevant and Bridges stayed on, supported by a grant from the Carnegie Institution of Washington. Muller was briefly a member of the Columbia department of zoology (1918–20), as were Donald Lancefield (1922–38) and Alfred Huettner (1922–28). Jack Schultz took Sturtevant's place on the Carnegie team when Sturtevant became a professor in 1929, and remained until 1942. Metz, who worked at the Carnegie Institution's laboratory at Cold Spring Harbor, was a member of the extended family—and a potential rival. These core drosophilists, together with Muller, Weinstein, and Altenburg (drosophilists in exile, as it were), constituted a *Drosophila* elite, bound by a shared sense of participating in a remarkable history.

The elite drosophilists were quite different from the rest of the thirty-odd students who passed through the fly room between 1910 and 1928 (see table 4.1). They did their dissertation research on problems central to the mapping project, such as linkage, chromosomal mechanics, crossing-over, nondisjunction: prestigious, productive problems. Rank-and-file drosophilists, in contrast, worked on the more biological and messier aspects of the *Drosophila* system. Gail Carver, for example, measured the variation of productivity and fertility in various kinds of mutants. John Gowen and Harold Plough studied crossing-over as a physiological process by measuring the effects of temperature, age, and other environmental variables on its rate. Don Warren studied the biometrics of egg size and its genetic and environmental determination. Clara Lynch and Shelly Safir worked on the genetic and physiological causes of sterility, and so on. Their work on the biology of *Drosophila* was important in practical ways for the mappers, but more in the way of control experiments. It was not fashionable, and it tended to bog down in biological complexities or turn up negative results. Francis Duncan won little kudos by proving that hybridization did not stimulate mutation, nor did Roscoe Hyde by showing that inbreeding did not cause sterility. Plough and Gowen's labors revealed little about the biological mechanism of crossing-over. It was a far cry from the elite's spectacularly productive work on linkage and crossing-over.

The *Drosophila* elite also had different careers from the rank and file, most of whom became teachers of general biology (table 4.1). Few first-generation drosophilists remained active researchers on *Drosophila*. Most published a few papers on their dissertation work and then settled down to teaching careers. A few who enjoyed some special local advantage did keep a specialized line of work going, as Harold Plough did at Amherst College, or John Gowen at the Rockefeller Institute, or Mary Stark, who pursued her interest in hereditary tumors in *Drosophila* at the New York Homeopathic Hospital. Not all the elite drosophilists had brilliant careers: Altenburg got stranded at Rice when Muller left, and Weinstein seemed quite unable to keep a steady job; but even so they never ceased to work at mainstream problems. It was not really until the late 1920s that rank-and-file graduates of the fly group could count on making careers as drosophilists.

Table 4.1
Fly Group Ph.D.'s, 1910–1930

Date	Student	Subject	Career	Papers
1910*	Quackenbush, Leopold	Sex ratio	Deceased	1
1913*	Rawls, Elizabeth	Sex ratio	—	1
1913*	Carver, Gail	Fertility	Mercer Coll	1
1914	Dexter, John	Genetic biology	Univ Puerto Rico	2
1914	Richards, Mildred	Development	—	4
1914	Sturtevant, Alfred	Chromosomal mechanics	Columbia	63
1914*	Tice, Sabra	Genetic biology	—	2
1915	Hyde, Roscoe	Fertility	Indiana Normal	15
1915	Muller, Hermann	Chromosomal mechanics	Texas, etc.	79
1916	Altenburg, Edgar	Linkage (*Primula*)	Rice	13
1916	Bridges, Calvin	Chromosomal mechanics	Columbia	83
1916	Metz, Charles	Cytogenetics (Diptera)	Cold Spring Har	32
1917	Gowen, John	Genetic biology	Rockefeller Inst	23
1917	McEwen, Robert	Physiology	Oberlin	2
1917	Plough, Harold	Chromosomal mechanics	Amherst	21
1917	Weinstein, Alexander	Chromosomal mechanics	Varied	12
1919	Lynch, Clara	Sterility	Rockefeller Inst	2
1919	Stark, Mary	Tumor mutant	NY Homeopathic	14
1920	Safir, Shelley	Chromosomal mechanics	City Coll NY	3
1922	Lancefield, Donald	Linkage (*D. obscura*)	Queens Coll	12
1923	Huettner, Alfred	Embryology	Queens Coll	13
1923	Warren, Don	Physiology	Kansas State	4
1926	Bliss, Chester	Development	Conn Agric Sta	3
1926	Plunkett, Charles	Development	New York Univ	4
1928	Bergner, Anna	Genetic biology	Cold Spring Har	1
1929	Harnly, Morris	Development	New York Univ	17
1929	Schultz, Jack	Development	Caltech	15
1930	Graubard, Mark	Physiology	U.S. Dept Agric	4

SOURCES: *American Men and Women of Science*, editions 1–7; and H. J. Muller, *Bibliography on the Genetics of Drosophila* (Edinburgh: Oliver & Boyd, 1939).

Date = Date of Ph.D. degree; * indicates M.A. or no degree

Subject = Dissertation topic

Career = Principal career position

Papers = Number of publications through 1938

Figure 4.1   The Columbia fly group in the chart room, celebrating Stur-
tevant's demobilization in 1918. *Left to right:* Edgar Anderson, Franz Schrader,
Alexander Weinstein, Alfred Sturtevant, S. C. Dellinger, Calvin Bridges, "hon-
ored guest," Alfred Huettner, H. J. Muller, T. H. Morgan, Otto Mohr, and
Frank Lutz. Courtesy of American Philosophical Society Library, Stern
Papers, photographs file Morgan.

Gender was another internal boundary. Except in its very early
years, the fly group was composed predominantly of men. Up to 1915
women were a majority of Morgan's recruits to *Drosophila* work, though
only one went on to an academic career. From 1914 to 1930, in contrast, only three women became drosophilists, along with eighteen men
(table 4.1). It looks as if women were attracted or pointed to *Drosophila*
when it was a marginal laboratory animal but not when *Drosophila* genetics had become a way of making mainstream academic careers. How
exactly the selection process worked is unclear; there is simply no evidence. We may surmise that women, who were not expected to have
mainstream careers, either felt unable to compete with men or were
steered to other lines of work. In any case, it is a fact that the fly group
after about 1915 was predominantly a male society. Women did work
in the fly lab. Edith Wallace, a trained biologist, served as the group's
artist in addition to being Morgan's personal technician.[1] Lillian Mor-

1. Edith Wallace employment form, 16 Feb. 1943, THM-CIT, file Biology
Division Salaries; Morgan to J. C. Merriam, 14 Dec. 1928, CIW; Morgan to

gan came back to experimental work about 1919, when her youngest child went off to school, and she carried out mainstream work. Her *attached-X* mutant, for example, was one of the group's most productive experimental systems in the 1920s.[2] Helen Redfield, who came to the group in 1925 as a National Research Fellow, became a permanent research fellow when she and Jack Schultz married in 1926. A spirited and independent woman, she remained "Miss Redfield," disdained gender stereotypes, and staked out her own line of research on the mechanism of crossing-over.[3] Wives of graduate students worked as technicians and stockkeepers. So the village society of the drosophilists was not monkishly male, but women did not occupy official positions; they were there as unpaid working wives and volunteers. They do not appear in official photographs. The group's formative psychosocial relationships were male: master and disciple, father and son, Boss and "boys."

The fly group in the Columbia years was thus divided between an elite, who were close to the central mapping project and had (or expected to have) careers as drosophilists, and a rank and file, who were more like ordinary experimental zoologists in their work and career expectations. Probably this division was in part a result of Morgan's laissez-faire attitude to graduate students; he allowed those who were ambitious and knew what they wanted to make a place for themselves in the new genetics, and he provided those who were less independent and aggressive with problems that suited his own more eclectic biological tastes and that prepared them for traditional academic jobs. In any case, the social division is a reminder of how peculiar *Drosophila* genetics was at first. We are accustomed to regarding the most famous drosophilists as the norm and the others as departures from the norm, but in fact the elite drosophilists were the oddballs in a social system that mainly produced general biologists for a collegiate market.

The peculiarity of Morgan's "boys" did not go unnoticed by other biologists. Ernest W. Lindstrom, a plant geneticist of just their age, was "somewhat disappointed" when he first met them at a scientific convention in 1914: "They are peculiar chaps, especially Bridges, Sturtevant and Muller. One wouldn't expect much from them on first sight, but

---

V. Bush, 1 Mar. 1940, CIW; and Theodosius Dobzhansky, oral history, Butler Library, Columbia University, New York, pp. 239–41.

2. Garland E. Allen, *Thomas Hunt Morgan: The Man and His Science* (Princeton: Princeton University Press, 1978), pp. 289–90, 295–95, 381.

3. H. Redfield to C. Stern, 14 July 1927, CS.

Hot Dog.

Figure 4.2 Alfred Sturtevant, 1918 or 1919—"Hot Dog." Courtesy of American Philosophical Society Library, Stern Papers, photographs file Sturtevant.

that means they are geniuses (?) I suppose."[4] Clearly, their reputation as cocky young stars had preceded them. In fact, the *Drosophila* elite were unusually ambitious and aggressive, and more devoted to a fast-paced, highly productive style of experimental work than was the norm. They worked in an ecological microenvironment that was different from ordinary academic departments of biology, including the department at Columbia. The core drosophilists quickly evolved a distinctive identity and culture of production, which reflected their peculiar symbiosis with the mapping project and the standard fly.

4. E. W. Lindstrom to L. J. Cole, 3 Feb. 1915, Archives of the College of Agriculture, University of Wisconsin, Madison, series 9/17/3, box 4. Quoted by Barbara A. Kimmelman, "Organisms and interests in scientific research: R. A. Emerson's claims for the unique contributions of agricultural genetics," in Adele E. Clarke and Joan H. Fujimura, *The Right Tools for the Job: At Work in Twentieth-Century Life Sciences* (Princeton: Princeton University Press, 1992), pp. 198–232, on p. 227.

The Fly Lab

Let us enter the workshop of the drosophilists as they would have, by climbing the stairs of Columbia's Schermerhorn Hall to the top floor and entering room 613. The "fly room" is small: just sixteen by twenty-three or twenty-six feet (accounts differ), about the size of the living room of a New York brownstone. It is crowded, crammed with eight desks and the paraphernalia of *Drosophila* production: odd-looking homemade incubators, shelves loaded with milk bottles, tables for preparing fly food. A large bunch of ripening bananas hangs in one corner. The place smells of yeast, fermenting bananas, and ether, and it is a mess—that made an indelible impression on all who entered the fly room for the first time, especially visitors from Europe like Otto and Tove Mohr in 1919 or Theodosius Dobzhansky in 1927. They expected a clean, well-appointed, all-American workspace and were taken aback by the jumble of bottles and equipment piled on the desks, the accumulated dust and dirt, and the swarms of cockroaches. Morgan's place was especially disgusting, owing to his habit of squashing flies on his counting plate (he never could get used to Bridges's tidy "fly morgue"), and legend had it that a rare cleaning of Sturtevant's desk disinterred a mummified mouse.[5] There were no personal, private spaces in the fly room, except for Morgan's small adjoining office, the door of which was always open. Space was arranged so that everyone knew what everyone else was up to.

The fly room was a noisy place, too, with the clinking of bottles and an unceasing flow of banter, gossip, and shoptalk. Every new result, or technical problem, or letter from another lab was openly and vigorously discussed by everyone present. "There was lively conversation going on," Tove Mohr recalled, "talking, joking, laughing—the professor in the midst of the group. They were like [a] bunch of students having a good time together." It had nothing like the formal hierarchy and decorum of European institutes, where professors were aloof and did not fraternize: "Here in the new world, it was like a family having everything in common." Jack Schultz recalled how Sturtevant "would look up from either his microscope or his clip-board, and begin (before you

5. T. H. Morgan, "Calvin Blackman Bridges," *Biog. Mem. Nat. Acad. Sci.,* 22 (1941), 31–48, on p. 33; A. H. Sturtevant, "Reminiscences of T. H. Morgan," 16 Aug. 1967, AHS 3.20; Tove Mohr, "Personal recollections of T. H. Morgan," n.d. [1961], AHS 3.18; Dobzhansky, oral history, pp. 239–41; Allen, *Morgan,* pp. 164–72, and Elof A. Carlson, *Genes, Radiation, and Society: The Life and Work of H. J. Muller,* (Ithaca, N.Y.: Cornell University Press, 1981), pp. 66–69, 75–76.

had opened your mouth) to tell you what he had just been finding out."[6] Morgan held "journal club" meetings at his home on most Friday evenings, but the real shoptalk was carried on in the shop. Dobzhansky recalled that Sturtevant "had a habit of talking much of the time as he was counting or examining the flies . . . or arranging crosses." It was a nonstop conversation, as Muller recalled: "'Oh,' Bridges would say, 'Here is a strange case I just got the results from.' We'd all discuss it, and I might make a suggestion about what to do next . . . and so, we simply went along in an informal way, talking things over with one another as they came along; very seldom on what you might call a philosophical plane."[7] In the Caltech years Schultz would spend hours every day talking to one person after another. Letters from colleagues were read aloud and avidly dissected—it was Sturtevant's "indoor sport," and everyone's.[8]

In the early years Morgan was the center of activity, as William Bateson noted on his visit in 1921. "Morgan supplies the excitement," he wrote his wife, Beatrice. "He is in a continual whirl—very active and inclined to be noisy." Sturtevant and Bridges, in contrast, were quiet. Bateson was pleasantly surprised: "[They] are quite different from the type I expected," he wrote. "Both are quiet, self-respecting young men. Sturtevant has more width or knowledge—Bridges scarcely leaves his microscope. I wonder whether they are not the real power in the place."[9] It was a canny observation, though clearly Morgan's boys were atypically well behaved for the great man's visit, seen but not heard. They were not usually so quiet, especially in the evenings, when Morgan was at home writing and the boys were taking the last of the three daily collections of virgin flies. "Bridges and I practically lived in this room," Sturtevant recalled. "[W]e slept and ate outside, but that was all. And we talked and talked and we argued, most of the time. I've often wondered since how any work at all got done with the amount of talking that went on, but things popped."[10]

6. T. Mohr, "Recollections"; and Schultz to Beadle, 31 July 1970, JS.

7. Dobzhansky, "Answers to W. Provine's questions," n.d. [Apr. 1975], TD file Mayr; and Muller, interview with G. E. Allen, March 1965, in Allen, *Morgan*, pp. 190–91.

8. Dobzhansky to Demerec, 19 Jan. 1935, MD; and Dobzhansky to Stern, 14 Jan. 1935, CS.

9. W. Bateson to C. B. Bateson, 26 Dec. 1921, Bateson Papers, microfilm ed.; and Allen, *Morgan*, p. 195.

10. Sturtevant, "Reminiscences of T. H. Morgan," 16 Aug. 1967, AHS 3.20; see also Carlson, *Muller*, pp. 75–76; and Allen, *Morgan*, pp. 194–95.

Figure 4.3 The Columbia Fly Room, with bananas, equipment for preparing
fly food, and general mess, circa 1920. Courtesy of American Philosophical
Society, Stern Papers, Photograph file Bridges.

The fly people were organized for production first and foremost,
and they produced collectively as a group and not as individuals working on personal projects. The overriding imperative was to get work
done, and credit went to those who got the work done first. The fly
people were hardly alone in this, of course, but they were more self-
conscious about their cooperative customs than most. Their collective
work style did not derive from some abstract moral ideal; it was an essential part of their commitment to production. That is what Jack
Schultz concluded, anyway, looking back on his student years:

> About the cooperative spirit in the Fly-lab. I've been thinking a good
> deal about that . . . and have now the feeling that it derives from Morgan,
> and paradoxically has not so much to do with cooperation as with the
> paramount importance attached to getting on with the work. I cannot
> recall any instance of explicit discussion of the value of cooperation; it
> was always taken for granted, and taught by example. When I decided to
> work with Minutes for my thesis, both Sturt and Bridges began saving
> new Minutes from their cultures for me: I had not asked them to do so,
> nor were they officially responsible for me in any way.[11]

11. Schultz to Beadle, 31 July 1970, JS.

Every publication that came from the fly group was to some extent a communal product, inspired and shaped by informal sharing, collaboration, and shoptalk.

The Fly Lab: Collective Training

The imperatives of production and collaboration also applied to the training of graduate students. Officially all graduate students worked with Morgan (he being the only professor), but in practice Sturtevant and Bridges also directed them. In fact, every graduate student was more or less a product of the whole group. Morgan, Sturtevant, and Bridges were not possessive of these apprentices and did not use them as hired hands on their own projects. They were junior members of the group, and all the core group took responsibility, in different ways, for their training and socialization. Schultz recalled his own experience:

> Morgan was my "major professor," In those days at Columbia [1924–28], Wilson was no longer accepting students, although [Robert H.] Bowen was beginning to do so—the alternatives in the department were not all that exciting. Moreover, the flow of post-doctoral fellows was continuous. Students were not a scarce item, and there was no particular need to be possessive about them.
>
> I suspect there is an underlying factor in this attitude, deriving from the way in which Dr. Morgan deputized the care and feeding of students. Mostly those who came to him were shunted to Bridges, who took care of the orphans. The effect of having such an arrangement was that official responsibility was diffused, and a sort of group responsibility came about. Bridges was one of oral examiners; but actually I considered myself more Sturt's student. In perspective, however, Dr. Morgan, whatever I may have thought at the time, really set the tone, and it was his example one aspired to follow, even though the specific day-to-day discussions were with the others. The pressure was to come up with one's own findings and ideas.
>
> . . . For [Sturtevant, students] were not commodities for one's aggrandizement, but people who were interesting to have around. There was no need to be aggressive about them. All Sturt did was be himself in his lab.[12]

Weinstein recalled that everyone participated in assigning dissertation problems to graduate students, especially Sturtevant and Bridges.[13]

12. Ibid. Cytologist Robert H. Bowen died in 1929 at the age of 37.

13. Weinstein to Sturtevant, 27 June 1957, AHS 3.19.

In fact, Morgan never liked assigning problems to graduate students and avoided it whenever he could. "Supplying graduate students with ideas and problems has exhausted me mentally," he complained to Edwin Conklin in 1907. "I have about as many as I can manage—students not ideas—and as you know it is something of a strain to fill up their deficiencies."[14] Students who asked Morgan to give them a problem were likely to be put off with a practical joke. Edgar Altenburg suffered from such a joke. Asked for a dissertation topic, Morgan took him across the hall into the aquarium room, dipped his finger into a stagnant tank, scrutinized the green liquid for a moment, then suggested that Altenburg might like to work on *Daphnia* (water fleas). To Altenburg, who wanted desperately to be a drosophilist, it seemed a calculated rebuff, and he decamped for the department of botany.[15] Intentionally or not, Morgan's dislike of shepherding students made the training of apprentice drosophilists a communal activity.

Similar customs and rituals can be seen in the way the group socialized postdoctoral visitors. Since Morgan did not like nursemaiding visitors any more than he did students, the task of assimilating newcomers usually fell to Sturtevant. Theodosius Dobzhansky found this out when, in response to his request for a genetics problem, Morgan tossed him an offprint of a paper by Chester Bliss that was all biophysics. Having just arrived all starry-eyed from Russia, Dobzhansky wondered if he had made a terrible mistake. However, Sturtevant quickly stepped in and gave him a simple project to get him started.[16] Sturtevant was the unofficial welcomer and socializer. He sat down with newcomers to learn of their special interests and craft skills, then suggested suitable projects. Bridges did the same with undergraduates in Morgan's advanced genetics course, using laboratory exercises "to get linkage data and other such chores."[17]

The fly group members were highly skilled in the rituals of group renewal, which taught greenhorns the basic tricks of the trade while contributing to the group's communal output. When the zoologist Curt Stern came to learn *Drosophila*, for example, Bridges and Sturtevant provided him with a series of apprentice projects, each more difficult and original than the last. First Stern did cookbook experiments, learning the basic techniques. Bridges then put him to work

14. Morgan to Conklin, Oct. 1907, EGC; see also Morgan to Mohr, 4 May 1920, THM-APS; and Allen, *Morgan*, pp. 200–201.

15. Carlson, *Muller*, pp. 63, 81.

16. Dobzhansky, oral history, pp. 243–45.

17. Schultz to Beadle, 31 July 1970, JS.

doing exact mapping of some genes that he had already mapped roughly, and they published the results jointly. A third project, on the genetics of variability, grew out of chance observations in his student prakticum. Sturtevant then set Stern to work on the effects of age and temperature on crossing-over, and that led him to make his first original and soon-to-be-famous work on crossing-over in the Y chromosome.[18] With Bridges and Sturtevant as his guides, Stern transformed himself from novice to star drosophilist in just two years. Dobzhansky had a similar experience (see chap. 8). Stern was unusually able and ambitious, but his rapid progress also reveals the unusual ability of the fly group to make the most of fresh talent.

The Fly Group: Moral Economy

The fly group's communal customs clearly made it productive and intellectually vigorous. But these customs also complicated the crucial process of assigning ownership of problems and credit for achievements. When every experiment was discussed by the whole group before, during, and after its performance, how could anyone possibly say who had what ideas? When everyone's project was to some extent a group project, how could individuals be given credit? The shoptalk that so stimulated the pace of production undermined traditional notions of ownership and credit, in ways that were potentially corrosive of group solidarity. Yet open fights over credit were surprisingly rare. How, then, were potential conflicts managed?

Many cases of intellectual debt were easily resolved by the universal rules of experimental science: coauthorship for work that was jointly done, public acknowledgment of borrowed tools or incidental suggestions. But what to do in a case where one person did the experiments but others contributed ideas fundamental to its success? In most groups such cases were rare, but in the fly group they were the rule. How did the fly group's moral economy cope with that?

It became the custom in such cases to give credit solely to the one who actually made the experiment work, even if others may also have conceived the idea. It was the drosophilists' golden rule. Ideas that were engendered in the group's informal shoptalk were treated as a communal resource and not as personal property. Such ideas could not be reserved and did not even need to be publicly acknowledged.

18. Stern, report on fellowship, to W. E. Tisdale 4 Oct. 1926, CS file IEB; and Sturtevant to O. Mohr, 12 Dec. 1924, 19 Dec. 1926, OM.

Figure 4.4 Alfred
Sturtevant, circa 1925.
Courtesy of American
Philosophical Society
Library, Stern Papers,
Photographs file Stur-
tevant.

In such a group it would have been divisive and disruptive of production to sort out credit for ideas, or to allow individuals to reserve problems, tools, or methods for their exclusive personal use.

The fly group never undertook to divide *Drosophila* genetics into
specialized subfields. Although most members of the group did have
favorite problems that they kept coming back to (usually several), they
never made these special lines the basis of their identity in the group.
Sex determination was Bridge's favorite, and Sturtevant was the expert
in *Drosophila* systematics and phylogeny, but many others also worked
on these problems from time to time. The internal divisions of labor
were not territorial but functional: technique, education, communication. Individual identities took shape around specialized talents for assimilating newcomers, trouble shooting, or talking shop. The system
was well adapted to a group that was organized to exploit the abundance of the breeder reactor, and that was half research institute and
half academic department.

Because the fly people had their own individual styles of work, however, they experienced the group's moral economy in different ways. Bridges, who turned up more mutants than anyone and more good projects than he could ever do himself, was totally unconcerned with credit.[19] Sturtevant, who was fecund of ideas and a steady, productive worker, never saw any reason to question the fairness of the system: everyone, he thought, "came out somewhere near even in this give-and-take, and it certainly accelerated the work."[20] Muller, in contrast, thought the system discriminated against people like himself, who were quick of wit but who did not, for one reason or another, work fast. He even came to believe that Morgan and Sturtevant had made the rules consciously to deprive him of his due: "[They] could do this by . . . establishing the rule that publication is all that counts in priority but that all ideas should be given freely by word of mouth and need not be acknowledged in print."[21] In fact, Muller just worked more slowly than Bridges and Sturtevant. He had a weakness for grand, spectacular experiments that took a very long time to do: mapping entire chromosomes at one go, for example, or measuring the natural rate of mutation (that experiment took ten years to perfect). No wonder Muller felt the system was unfair.

The fly group's collective work style and moral economy were consequences of everyone's working on a single, highly productive laboratory creature and on problems that had grown out of the mapping project. Indeed, the fly group of the 1920s and 1930s was in many ways simply an expanded and diversified form of the original mapping project that Morgan and his young students embarked upon in March 1912. The programmatic vision was there from the start, and the division of labor, and Morgan's willingness to delegate control of the work to his students. By 1930, of course, the "boys" were no longer students but world-famous scientists, more than Morgan's equal in *Drosophila* know-how, though not his equal in academic rank. The peculiarities of the fly group's productive and moral economy derive, I think, from the way in which the structure of the group remained the same even as its core members moved through the career cycle. It was unusual for graduate students to collaborate in a dissertation project. And it was highly unusual for such a project to become permanently installed as

19. Bridges to Stern, 1 Feb. 1928, CS.

20. Sturtevant, untitled lecture, n.d., AHS 3.20; and A. H. Sturtevant, "Thomas Hunt Morgan," *Biog. Mem. Nat. Acad. Sci.* 33 (1959): 283–325, on pp. 294–98.

21. Muller to G. Allen, 7 July 1966, HJM-A file Altenburg.

a fixture in an academic department. How, then, was the mapping project perpetuated, and along with it the ambiguous relation between the Boss and his boys?

The Carnegie Grant

It was Carnegie money—grants-in-aid from 1915 to 1918 and a program grant from 1919 to 1942—that stabilized the distinctive microecology of the fly group. It was not just the money, however, but also the way it was understood and controlled by Morgan. It is striking how uneasy Morgan seemed to be with third-party funding. Many people have remarked on the contrast between his parsimony with Carnegie's money and his generosity with his own.[22] Morgan seemed to regard a grant as a contractual obligation, to be entered into cautiously and doled out grudgingly. Amazingly, he did not seek extramural funding for the mapping project until 1914, even though he could have used it much earlier. As early as March 1911 Davenport offered to intercede on his behalf with the Carnegie Institution, and it is perfectly clear that Morgan could have had a grant simply for the asking. Yet Morgan declined to apply, even though he was barely able to cope with the flood of mutants; he even apologized for sounding "fearfully self centered and self conscious," as if asking for support was a personal weakness.[23] Why was he so reluctant? Morgan later suggested it arose from a fear of promising more than he could deliver. "I decided to ask for assistance," he later told CIW president Robert S. Woodward, "only when I felt quite certain that the work had gone so far that there could be no question about our being able to make some return."[24] Evidently Morgan felt uneasy about being accountable to an outsider.

Extramural support did entail a greater sharing of control than academics were used to. University fellowships and work-study jobs entailed minimal obligations to produce research, because research was an incidental by-product of credentialing. The Carnegie Institution, in contrast, expressly forbade recipients of grants-in-aid from using CIW

22. Sturtevant, "Morgan," p. 298; Dobzhansky, oral history, pp. 248–51; Muller to J. Huxley, 22 Nov. 1920, HJM-A; and Morgan to Merriam, 26 Sept. 1923, 26 Apr 1928, CIW.

23. Morgan to C. B. Davenport, 14 Mar. 1911, CBD; Davenport to Morgan, 10 Mar. 1911, CBD; and Allen, *Morgan*, pp. 249–57.

24. Morgan to Woodward, 17 Dec. (quotation), 21 May 1914, CIW.

funds to help students get degrees. The money was for individuals to produce work. Woodward had for years been fighting openly with his academic clients, trying to make them understand that using grant funds for student workers was an abuse.[25] Morgan certainly knew of Woodward's bête noire—who then did not?—and I suspect he preferred to make do with student aid and keep his academic freedom of action.

What made Morgan finally seek Carnegie support in 1914 was the immanent prospect of losing university funding of his gene mappers. Sturtevant received his Ph.D. in 1914, Muller in 1915, and Bridges in 1916, and Metz went to work full-time at Cold Spring Harbor in 1914. Sturtevant and Bridges could no longer be supported intramurally, and the mapping project was far from finished. Morgan could in theory have trained a second generation of students, but that would have meant serious delays in production, and Sturtevant and Bridges would have been formidable competitors of whoever succeeded them at Columbia. The *Drosophila* project had already become a project of the kind that the CIW supported in its in-house departments. It had outgrown small-scale, improvised academic funding. The impending breakup of the mapping team drove Morgan to accept outside support, with its uncertain moral obligations. He talked to Woodward in April 1914 and applied formally in May. In December the CIW's trustees approved a three-year annual grant of $1,800 for Sturtevant and $1,700 for Bridges to complete the chromosome mapping project, since Morgan thought they could more or less finish the job in three years.[26] His projection completely failed to anticipate the autocatalytic character of *Drosophila* production, and in 1917 he had to work up his courage to ask Woodward for a renewal. Woodward was of course eager to renew; few of the CIW's grants-in-aid were as novel and productive.[27] The fly group thus slipped into a permanent dependence on their patron.

Carnegie funding of the fly group was put formally on a permanent basis in 1919, when Sturtevant and Bridges received official appointments as "associates" (i.e., extramural employees) of the Carnegie

25. Robert E. Kohler, *Partners in Science: Foundations and Natural Scientists, 1900–1945* (Chicago: University of Chicago Press, 1991), chap. 2.

26. Morgan to Woodward, 21 May, 17 Dec. 1914, 1 June 1915; and Woodward to Morgan, 23 May, 14 Dec. 1914; all in CIW.

27. Morgan to Woodward, 24 Sept., 2 Oct. 1917; and Woodward to Morgan, 26 Sept., 4 Oct., 29, Dec. 1917; all in CIW.

Institution with effective tenure and full pension rights.[28] As was the CIW's custom, it gave the money to Morgan personally, not via Columbia University, so that there would be no doubt that Morgan was personally responsible for spending the money to good effect. The fly group thus became in effect an extramural department of the Carnegie institution, encysted in Columbia's department of zoology but largely independent of it. And thus did Sturtevant and Bridges become researchers who enjoyed complete freedom in their work but who were very much dependent on the Boss for their salaries and material support for their work.

What impelled Morgan into a closer relation with the CIW was again, as with the 1914 grant-in-aid, the threat of losing his star drosophilists. This time the threat came from within the CIW itself, from its genetics lab at Cold Spring Harbor and Charles Metz. So long as Metz was a one-man show, he posed no threat to the Columbia group, but it was quite a different matter when in 1919 he won the backing of Davenport and Woodward to create a fly group of his own—with a budget of no less than $15,000 a year and a "life-sized" program of research in comparative and evolutionary genetics. Metz even talked of luring Sturtevant and perhaps Bridges to Cold Spring Harbor with the prospect of more ample material support than Morgan was giving them, hinting that the *Drosophila* world might be divided, with the Columbia group taking *D. melanogaster* and Cold Spring Harbor all other species of *Drosophila*.[29] Morgan stood to lose not only his best drosophilists, but perhaps also his Carnegie grant, should the CIW decide that it was inefficient to fund two competing and adjacent centers of *Drosophila* work.

Embarrassed that Metz had proved the more vigorous and foresighted grant getter, Morgan quickly sent Woodward his own competing scheme for an expanded project. Metz, for his pains, received a sharp paternal lecture on the ethics of cooperation. If he did not understand that any restriction on the free use of *Drosophila* was harmful

28. The CIW also paid for Edith Wallace and a full-time stockkeeper, bringing the total grant to about $10,000 per year: Davenport to Morgan, 3 Mar. 1919, CBD; Morgan, "Memorandum relating to suggested changes in the work on heredity of *Drosophila*," 13 Mar. 1919, CIW; Woodward to Morgan, 15 Mar., 16 Apr. 1919, and Morgan to Woodward, 23 Apr. 1919, CIW; Allen, *Morgan*, pp. 251–57; and Morgan to O. Mohr, 6 Mar. 1919, THM-APS.

29. C. Metz, "Brief statement of scientific program for Drosophila investigations," n.d. [Jan. 1919], CBD; and Metz to Davenport, 14 Nov., 8 Dec. 1918, 27 Jan. 1919, CBD. Metz may have suggested a division of labor to persuade Woodward, a rationalizer, that funding both centers was efficient (see below, p. 121).

to all drosophilists, Morgan chided his wayward disciple, then "the import of the teaching in the Columbia laboratory has failed to take root, viz. that the best work can be carried out by constant cooperation between individuals working on the best materials obtainable, and that the moment they begin to appropriate special fields of influence, the work will suffer." Privately, Morgan bragged that Metz could never put together a team as good as his.[30]

In fact, Woodward was only too glad to expand two such successful groups, though it took some negotiation to work out what their relation would be. "We almost had a grand scrap before all that was straightened out," Sturtevant reported to Otto Mohr. "Then after the scrap was averted, I was scared for a while that the whole bunch of us might move from here, and lose all connection with any university, the Boss becoming a Carnegie research man." In the end, arrangements were made for the rival groups to cooperate, and rooms were renovated in Schermerhorn Hall for Metz, José Nonidez, Rebecca Lancefield, and a technician. "Not a poor group," Sturtevant thought, "but hardly the lot of first-class, highly trained geneticists that he [Metz] started out to get together."[31] To keep the peace, Woodward enjoined Davenport and Metz from ever again making offers to anyone in the CIW's employ. The fly group was off limits: no poaching.[32]

Permanent funding guaranteed the fly group's continuing existence as a quasi research institute, although Morgan never ceased to fear that the CIW might one day decide to consolidate the two groups at Cold Spring Harbor. The CIW's grant also perpetuated established patterns of authority and control within the fly group. Sturtevant and Bridges controlled the experimental work, but Morgan alone dealt with the CIW and controlled the allocation of CIW funds—down to the last penny. This disparity in status and control caused tensions, as we will see, when Sturtevant and Bridges reached a stage in their careers when they would normally have enjoyed institutional powers. Morgan's parsimony in doling out CIW funds became a chronic irritant in the fly group. While CIW funding freed the group from the restric-

30. Morgan to Metz, 26 Feb. 1919, THM-APS; and Morgan to Mohr, 6 Mar. 1919, THM-APS. See also Allen, *Morgan*, pp. 252–54.

31. Sturtevant to Mohr, 30 Mar., 25 Sept. 1919, OM.

32. CIW executive committee memo, 14 Feb. 1919, Metz to Davenport, 22 Feb. 1919, Davenport to Metz, 3 Mar. 1919, Davenport to Morgan, 9 Jan., 7 Apr. 1920, Morgan to Davenport, 24 Jan. 1920, all in CBD; and Morgan to Woodward, 14 June, 29 Aug. 1919, Morgan to Merriam, 22 Jan. 1924, all in CIW.

tions that normally limited the size and life span of student projects, it also perpetuated an anomalous moral relationship, in which the Boss's most accomplished students remained perpetually his boys.

Boss and Boys

There are many little hints that Morgan and his boys never quite outgrew the habits of teacher and students. In 1926, for example, Sturtevant abandoned his plans to travel to Berlin when Morgan did not encourage it. "I don't know why T.H.M. should be mixing in on Sturt's plans like that," his friend Franz Schrader told Stern, "but I suppose that any request to the Carnegie for travelling would have to go through him."[33] The same thing happened a year later, when Sturtevant wanted to attend an international congress but resigned himself to staying home when he noticed that Morgan had allocated no money for it in the budget: "In the absence of any further information from the Boss," he wrote Stern, "that ends the matter."[33] In 1928 an invitation to Bridges to visit Berlin was sent to Morgan, who seems to have lost the letter; Bridges only learned of it by chance from friends. In 1929 Sturtevant wanted the group to have a technical assistant in cytology, but Morgan did not approve and Sturtevant felt he could not press him.[34] Clearly, Morgan would never have presumed to direct his chief drosophilists on matters of experiment; in that domain their greater skill gave them nearly absolute authority. But in matters of money and politics Morgan had all the status and authority; there he was boss, and the boys deferred.

Sturtevant had the most complex and ambivalent relationship with the Boss. He was Morgan's favorite and was like Morgan in many ways: in his family background (old New England stock that had moved West and South), his Anglophilia, and his broad biological interests. He wrote on the heredity of racehorses, helped Morgan sort out the heredity of mice and sheep, and once planned to write a book on heredity and animal husbandry. He was an avid natural historian and taxonomist and collected bugs whenever he was in the countryside.[35] Sturte-

33. Schrader to C. Stern, 5 Feb. 1926, CS; and Sturtevant to Stern, 8 Jan. 1927, CS.

34. Bridges to Stern, 1 Feb. 1928, CS; and Sturtevant to Stern, 23 Mar. 1929, CS.

35. Sturtevant, "AH Sturtevant biographical notes," n.d., AHS 4.2; G. W. Beadle, "Alfred Henry Sturtevant (1891–1970), *Amer. Phil. Soc. Yearbook* 1970: 166–71; and E. B. Lewis, "Alfred H. Sturtevant," *DSB* 13: 133–38.

Figure 4.5 Caricature of Alfred Sturtevant by Curt Stern, 1925 or 1926. Courtesy of American Philosophical Society Library, Stern Papers, small memo book in box of unsorted notebooks.

vant was always the one most privy to the Boss's plans and was always regarded as his heir apparent.

Sturtevant evidently constructed his professional identity by observing and emulating Morgan. At professional meetings he was drawn to the smoke-filled rooms where Morgan and his cronies gossiped, argued, and engaged in a bantering but pointed repartee about the new genetics.[36] Sturtevant adopted the same style of high-spirited but sharp-edged banter. A letter to Roy Clausen from 1921 is characteristic:

> Eugenics conference wasn't so worse. The long-haired guys weren't so numerous as expected, and flocked mostly by themselves. A partial list

36. Sturtevant, "Personal recollections of T. H. Morgan," 25 Apr. 1966, AHS 3.20. For examples of this bantering style see Morgan to Conklin, 8 June 1905, 25 Nov. 1925; and Conklin to Morgan 10 June 1905, 30 Nov. 1925; all in EGC.

of the regular, or nearly regular, guys I ran around with follows. Gowen, East, Anderson, Jennings, E. N. Harvey, W. K. Fisher, F. E. Lutz, A. F. Shull, G. H. Shull. Etc. etc. . . . One interesting affair came off when I heard G. H. Shull and Anderson get together for the first time, and [I] kept them going on Oenothera for an hour or so. Oh, man, you should have been there!"[37]

Sturtevant especially admired Morgan's productivity and range of experimental interests. "Think of getting out 12 papers in one year, or 42 in four years," he exclaimed to Mohr, "and on such a variety of subjects! Some Boss, I'll say!"[38] In his early and middle years Sturtevant was a confident indeed cocky, young star. In his maturity he was the fly group's sage and prima donna. "Very matter of fact, very self-possessed," Dobzhansky recalled, "a steady worker, no flashes, no sparks, just good steady work everyday, including Christmas Day, New Year's Day, always in the laboratory, always counting flies."[39]

Sturtevant was less adept at picking up Morgan's tact and politeness. He was careless of injuries inflicted by his habit of blunt criticism, and he sometimes failed to see when high spirits became rudeness and intolerance. Morgan was well aware of his protégé's bad habits. When Sturtevant married Morgan's technician, Phoebe Reed, in 1922, Morgan hoped that his new wife would "help to make him a little more civilized. . . . Already he shows signs of a little more consideration for other people, and I hope he will form the habit, for, at bottom he is a fine, straight fellow, a great deal better than his manners."[40] Sturtevant's manners never entirely lost their rough edge, however. He could turn unexpectedly even on very close friends like Theodosius Dobzhansky, as we will see. With most people Sturtevant was friendly and helpful, but with those of whom he disapproved he could be intolerant and mocking, as he was with the gentle and able Hungarian cytologist Peo Koller, apparently because Koller had as a young man been a Benedictine monk.[41] Sturtevant was not a mean man, just intellectually arrogant, excessively high-spirited, and not always in control of a sharp-edged sense of humor.

37. Sturtevant to R. E. Clausen, 11 Oct. 1921, 12 Jan. 1931, UCDG.

38. Sturtevant to Mohr, 3 Dec. 1922, OM.

39. Dobzhansky, oral history, pp. 268–69.

40. Morgan to O. Mohr, 16 Mar. 1922, THM-APS.

41. Dobzhansky to P. D. Dansereau, 18 Apr. 1968, TD; Dobzhansky, oral history, pp. 436–39; Peo Koller to Muller, 8 Mar., 13 July 1938, HJM-C; and A. H. Carson to W. Provine, 19 Feb. 1980, WP.

If Sturtevant cultivated the image of an intellectual aristocrat, Bridges was drawn to the quite different role of technical expert and worker. He was the fly group's self-styled blue-collar worker. Orphaned as a young child. Bridges was raised by puritanical grandparents in a small town in upstate New York and grew accustomed to earning his keep from an early age. He was a scholarship student at Columbia and made money in any way he could, including selling books door to door. A country boy in the big city, Bridges put on some New York sophistication, such as a belief in communism and "free love," but he never lost the naïveté of his small-town upbringing. He was unaffected, strikingly beautiful, gullible, and quite indifferent to social conventions and appearances. He was instinctively tolerant and generous. He never refused to help a student or colleague, never held a grudge, and never engaged in a dispute over priority.[42] When Richard Goldschmidt laid claim to his theory of sex determination, Bridges refused to argue: "Tell him," he wrote Stern, "I don't claim to be very original, only to blunder along interpreting my own data and developing what ideas come to hand in that connection. We all do that."[43]

Bridges was not a great talker or gossip but took boyish delight in swapping *Drosophila* lore, when he could be bullied or shamed into answering letters. "Caught at last—too many questions to be ignored!" began one letter to Curt Stern. Or, to Otto Mohr, "'At last! that confounded good-for-nothing Bridges has condescended to answer my urgent letters—now that the information is too late to do me any good.' Thus Otto." Hot news was announced effervescently in a large, boyish scrawl: "Oh Boy! Got a new Y that carries an ordinary bobbed allele."[44] Bridges worked in fits and starts, puttering about for long periods and procrastinating, then working like one possessed, brilliantly. He had, as Dobzhansky put it, a divine spark, or *Götterfunken*.[45] He fascinated and charmed everyone who met him.

Bridges was famous, too, for his flamboyant private lifestyle. He left his wife and children in the early or mid 1920s (though continuing to support them), got a bachelor pad and a vasectomy, brewed dreadful moonshine, and took to practicing "free love," which meant propositioning every woman he met, indiscriminately. He took refusal matter-

42. Morgan, "Bridges" (see n. 5). The unpublished draft of this biographical memoir in THM-APS is richer in detail.

43. Bridges to Stern, 13 Jan. 1927, CS.

44. Bridges to Stern, 13 Jan. 1927 (quotation), 1 Feb. 1928, CS; Bridges to Mohr, 13 Mar. 1923, OM; and Dobzhansky to Demerec, 19 Jan. 1935, MD.

45. Dobzhansky, oral history, pp. 260–62.

Figure 4.6   Calvin Bridges at work in the Columbia fly lab, early 1920s. Note the working chromosome map directly behind him. Courtesy of American Philosophical Society Library, Stern Papers, Photographs file Bridges.

of-factly, but his extraordinary beauty and innocence were irresistible to many. Sometimes Bridges's hobby got him into trouble, as in 1933, when a Harlem confidence woman, posing as an Indian princess, picked him up on the train to Woods Hole and persuaded him that her father was a maharaja and wanted to set him up as head of an endowed institute of *Drosophila* genetics in India. (Bridges was so taken in that he offered Schultz a job in his institute!) Kept in high style at Woods Hole, the "princess" followed Bridges to Pasadena as "Mrs. Bridges," leaving debts in her wake and threatening to prosecute Bridges for transporting her across state lines for immoral purposes. In the end Morgan scared her off by threatening to call the police and sent Bridges into hiding for a few months.[46] Asked by Milislav Demerec to refrain from having affairs with the Cold Spring Harbor staff during a forthcoming visit, Bridges cheerfully agreed, commenting that design-

46. Ibid., pp. 262–68; Schultz to Sturtevant, 8 Aug. 1966, AHS 5.13; and Bridges to Mohr, 12 Dec. 1933, OM. Bridges referred to "family complications" in Bridges to Mohr, 5 Dec. 1923, OM.

ing and building cars—his other off-hour obsession—was "a hobby that's much safer than women."[47]

Communism and free love were not incongruent aspects of the professional identity that Bridges constructed for himself, that of a scientific worker. His hobbies disqualified him for an academic or administrative position, and that may even have been part of their appeal. Where Sturtevant was loaded with official honors, Bridges had to be content with the unofficial appreciation of working drosophilists. Morgan liked and deeply respected him but was never as close emotionally as he was to Sturtevant; they were just too different in personality, politics, and lifestyle. Their relationship had ups and downs: "Bridges has calmed down somewhat, and learned a bit of discretion," Sturtevant reported prematurely to Mohr in 1931, "so he gets along much better with Dr. Morgan."[48]

Sturtevant and Bridges together embody the fly group's dual character as research institute and academic department. Sturtevant behaved like a research professor, Bridges like an employee. Sturtevant ranged widely in his intellectual interests, while Bridges, like a skilled worker, stuck to the Carnegie stocks. Sturtevant, as Muller recalled, was "the reader for the whole Drosophila group."[49] He was the master of the literature, keeper of the reprint collection, walking bibliography, and sage. Bridges, in contrast, was the group's technical expert and keeper of the stocks. Of 365 mutants listed in 1925, about 240 or two-thirds were found by Bridges.[50] As Muller observed, "Bridges . . . just runs across things because he examines as many flies and runs as many crosses in which he makes counts, as everyone else put together."[51] Sturtevant had no formal routine tasks in connection with the Carnegie grant. Bridges had many, and when he put them off too long, as he often did, Morgan would nag and bully him into finishing, as an employer might an employee.[52]

47. Demerec to Bridges, 21 Mar. 1934, MD; Bridges to Demerec, 26 Mar. 1934, MD (quotation); and Bridges to Mohr, 12 Dec. 1933, OM.

48. Sturtevant to Mohr, 11 Feb. 1931, OM.

49. Muller to Sturtevant, 19 Nov. 1946, HJM-A.

50. T. H. Morgan, C. B. Bridges, and A. H. Sturtevant, *The Genetics of Drosophila,* (The Hague: Nihoff, 1925), pp. 240–58.

51. Muller to J. Huxley, 12 May 1921, HJM-A.

52. Morgan to Mohr, 6 Sept. [1921 or 1927] (but see also 4 May 1921), THM-APS; and Bridges to Mohr, 5 Dec. 1923, OM.

Figure 4.7   Calvin Bridges,
1922. Courtesy of American
Philosophical Society Library,
Stern Papers, Photographs file
Bridges.

Sturtevant and Bridges complemented each other in the fly group's productive economy. As Sturtevant saw to it that students and visitors had projects, so Bridges served as general troubleshooter and ever-ready source of technical know-how. As Sturtevant put it, "He is . . . as indispensable to the rest of us as ever, still the final source of information as to all the mutant types and all the stocks, and still making up grand new stocks for the rest of us to use." Jack Schultz realized after Bridges's death how much he had unconsciously depended on him: "I am constantly catching myself up," he wrote Sturtevant, "thinking that after all it might be better to leave such and such a job until he [Bridges] got around to doing it." Dobzhansky remarked that he would have hated to be saddled with all the routine maintenance of the *Drosophila* stocks that Bridges did, and he was grateful that Bridges did it and not he.[53] Sturtevant generated ideas and inspired some of

53. Sturtevant to Mohr, 11 Feb. 1931, OM; Schultz to Sturtevant, n.d. [1939], JS; and Dobzhansky to Stern, 27 Jan. 1939, CS.

the group's major experimental lines (more on that later). Most of the material technology of *Drosophila* genetics was constructed by Bridges. He was always tinkering, always improving, like a production engineer, keeping the production line running smoothly. Bridges was not, however, the supertechnician that some thought him.[54] Like a master engineer, he understood *Drosophila* better than anyone; it was, after all, largely his creation.

Insiders and Outsiders

Tensions are especially revealing of a group's productive and moral economies, because they indicate important social boundaries: between one generation and another, insiders and outsides, privileged elites and rank and file.

The drosophilists' meteoric ascent was watched with mixed feelings by other groups in Columbia's department of zoology. The paleontologist William Gregory later recalled how the new genetics drew graduate students away from his own discipline. Although Morgan never put impediments in the way of those who chose to follow more traditional ways, Gregory resented having to live in the long shadow of the fly.[55] Gary Calkins was another member of the zoology department who felt eclipsed by Morgan and his acolytes, as became apparent when the fly group left for California in 1928, leaving Donald Lancefield and Alfred Huettner behind. Franz Schrader described the scene:

> I think it is a safe gamble that he [Calkins] will use all his power—and that is not small at Columbia—to get all the threads into his hands. . . . What poor Lancefield is going to do, is not clear to anybody—least of all to himself. But it seems certain that none of the former adherents of Morgan can expect any great favor for a while. . . . Huettner also feels very downcast by the whole thing and fears that Calkins will not make it easy for him over at Long Island. The worst of it for both Lancefield and Huettner is the apparent opinion of both the Boss and Sturt that neither is entitled to a better fate than they are getting. I fear that that may be right, but from the personal standpoint it seems very sad.[56]

54. John Bonner to W. Provine, 24 Aug. 1979, WP.

55. W. K. Gregory, "Fifty years in the department of zoology, 1896–1946," n.d. [May 1946], LCD.

56. Schrader to Stern, 29 Oct. 1927 (see also 5 Feb. 1926), CS.

Calkins did take revenge, vetoing Morgan's suggestions for new appointments and promoting the mouse geneticist L. C. Dunn over Lancefield's head—"a clever way of making things disagreeable for Lancefield," Schrader assumed.[57]

The fly group's most famous conflict was in the family, however: between H. J. Muller and Morgan and Sturtevant, who Muller believed had taken credit for achievements that were rightfully his.[58] There is no simple explanation of Muller's resentment. His hypersensitive and combative personality was clearly part of it, and there may also be something in Garland Allen's suggestion of class and ethnic differences, though it should be noted that Muller's Jewishness and working-class consciousness were very much an assumed, constructed identity.[59] More important, I think, is Muller's position in the working order of the early fly group.

The key point is that Muller was *not* present when the mapping project was designed and parceled out in March 1912. Although Muller had been in and out of Morgan's lab in 1910–12, he did not have his own desk in the fly room until September 1912, by which time Sturtevant and Bridges had already divided up mutants and chromosomes between them.[60] A few months made all the difference. Muller was not deliberately excluded from the project, though he came to think so; it was simply an accident that he had not been available for full-time fly work at the moment when the mapping group was formed.

Left on the fringes of the mapping project, Muller proceeded to construct a role and an identity for himself that emphasized the difference between himself and the core group but at the same time challenged them. He took up problems that were related to the main pro-

57. Schrader to Stern, 7 June 1928 (but see also 13 Nov. 1928, 14 Mar. 1931), CS. On Huettner see Schrader to Stern, 14 Mar. 1931, CS. On Lancefield see J. A. Moore to Provine, 11 Sept. 1979; Provine to D. Lancefield, 26 Dec. 1979; and Provine, notes on talk with R. Boche, 20 Dec. 1979; all in WP.

58. Carlson, *Muller,* pp. 79–89; Allen, *Morgan,* pp. 202–208; Elof A. Carlson, "The *Drosophila* group: The transition from the Mendelian unit to the individual gene," *J. Hist. Biol.* 7 (1974): 31–48, on pp. 39–47; and Muller to Sturtevant, 30 Apr. 1959, HJM-A.

59. His mother's family did trace their roots to Sephardic Jews, but they had been assimilated in England for generations and boasted an Archbishop and a marquis. The Mullers were German-American artisan manufacturers, well-educated, cultured, Unitarian, urban bourgeois. Carlson, *"Drosophila* group," pp. 10–22.

60. *Ibid.,* pp. 37, 51, 57–58; and Muller to Sturtevant, 20 Apr. 1959, HJM-A.

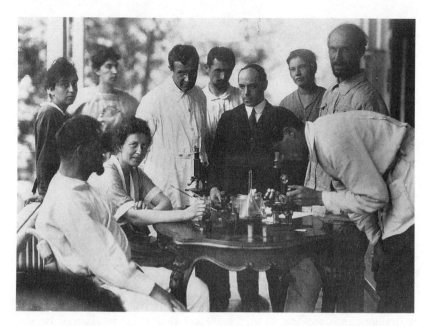

Figure 4.8   Hermann J. Muller with first generation of Russian drosophilists, Moscow 1922. Courtesy of American Philosophical Society Library, Michael Lerner Photograph Collection.

gram of linkage and crossing over, but in a way that was calculated, consciously or not, to demonstrate that he could beat the mappers at their own game: for example, in analyzing multigenic systems like *beaded,* or in the fashioning of elaborate mapping stocks:

> I didn't want to butt in and step on their toes, since they already had this linkage story going; so I decided on some more out of the way things to do. . . . They went ahead with their mapping, and I had the fortune of finding the gene on the fourth chromosome which completed the map for them. . . . I was on the outskirts of most of that because, while I was discussing it with them, I was not actually working with it in my hands.[61]

Muller thus contributed marginally to a project in which he was unable to participate directly and share officially in the credit. No wonder he came to resent the group's custom of taking ideas as a common good. Alienation came gradually: Schultz recalled Sturtevant's telling him

61. Muller interview, in Allen, *Morgan,* pp. 204–205; see also Muller to Allen, 7 July 1966, HJM-A file Altenburg.

how "in the early days Muller at first would say that he had developed such [and such] an idea independently, but wound up by saying that he had thought of it first.[62]

As Sturtevant and Bridge's deference to the Boss was a vestige of the way that the fly group was constructed, so, too, was Muller's undying resentment. Significantly, Muller's resentment did not extend to Bridges, in part because they had the same political beliefs, but also because Muller saw Bridges less as Morgan's protégé than as an exploited worker. Bridges did benefit from the group's system of giving credit, but Muller forgave him because he assumed that Bridges had to play along in order to keep his job![63] Muller regretted that they never worked together after Muller left the group.[64]

The dynamic of Muller's relation to the fly group might have been different had Muller become a permanent member of the Columbia department in 1920. Muller wanted every much to stay, and when his promotion was turned down he was convinced that Morgan had blackballed him. Since Morgan was in California at the time, it was more likely the chairman of the department, E. B. Wilson, who decided that Muller was more trouble than he was worth. Morgan was tolerant of difficult people so long as they were good at their work, and he was generous in acknowledging Muller's real achievements.[65]

Departure from the group was not always so traumatic. Charles Metz, for example, was no less ambitious than Muller but a good deal more self-assured and independent. Hanging around and challenging the boss was not for Metz; he just staked out an independent line of work and an independent career, going to work full-time at Cold Spring Harbor two years before finishing his degree. Perhaps it was his upbringing in the frontier town of Sundance, Wyoming, that made him so striving and independent, or the year he worked as assistant to Stanford's president David Starr Jordan. In any case, his energy got people's attention. "He is *pushing* and masterful and will doubtless have a tendency to dominate any situation into which he may enter," George Shull wrote of him. "He has more of the western *honesty* approaching

62. Schultz to Carlson, 7 Dec. 1967, JS.

63. Muller to Allen, 7 July 1966, HJM-A file Altenburg.

64. Muller to Bridges, 3 Jan. 1939; and Bridges to Muller, 3 May 1937, 1 June 1938; all in HJM-C.

65. Sturtevant to T. Sonneborn, 5 May 1967, AHS 3.19; Sonneborn to Carlson, 2 May 1967, AHS 3.19; Carlson, *Muller,* pp. 116–17, 120–21; Morgan to E. E. Bogart, 9 May 1928, HJM-C; and Muller to J. Huxley, 28 June, 22 Nov. 1920, 12 May 1921, HJM-A.

to brusqueness, than of the Southern *suaveness* exhibited by Dr. Gee [a competing applicant], so that you might find him a little less agreeable to work with unless sturdy straight-forwardness makes a special appeal to you."[66] In fact, Davenport had been pressing Metz to join his group for nearly a year.[67]

At Cold Spring Harbor Metz remained a member of the fly group's extended family, collaborating intensively with Sturtevant on the comparative genetics of *Drosophila,* until the dustup over CIW funding in 1919. It is unlikely that Metz meant simply to raid Morgan's group. Probably he thought a division of the *Drosophila* field between the two CIW labs would persuade CIW president Woodward that there would be no duplication of effort, not stopping to think how threatened Morgan would be by such a scheme.[68] In any case, Morgan's violent reaction and enduring suspicion of Metz's intentions stand to remind us of the potential for generational conflict that was built into the drosophilists' peculiar history.

Generations

It was not just Muller and his friends who harbored ambivalent feelings about Morgan. Bridges and Sturtevant did too, especially Sturtevant, as Dobzhansky recalled:

> It was literally true that each of the three—in this case I better talk about Bridges and Sturtevant, whom I saw virtually every day—had this intense resentment and jealousy, largely to Morgan, and also to each other. It was unavoidable that very soon I was . . . sitting with Sturtevant, and this conversation about Morgan having taken the credit for what has been made by him, Sturtevant, was repeated again and again and again and again. One of the quotations which he ascribed to . . . E. B. Wilson . . . was that "Morgan has made one and only one important discovery in his life; that discovery was Sturtevant."[69]

Dobzhansky's memory is not always to be trusted, but corroborating evidence of his account is found in letters that Franz Schrader wrote to Curt Stern at the time of the fly group's migration to Pasadena in

66. G. H. Shull to Babcock, 14 Nov. 1913, UCDG.

67. Metz to Davenport, 4 Feb. 1912, 4 Oct., 20 Dec. 1913, CBD.

68. Metz to Davenport, 14 Nov., 8 Dec. 1918, 27 Jan. 1919, CBD.

69. Dobzhansky, oral history, pp. 270–72.

1928. It was the disruption caused by this move, in fact, that caused latent tensions to surface.

As part of Morgan's plan for the new Biology Division at Caltech, Sturtevant gave up his job as Carnegie "associate" and became a professor and director of the division's department of genetics. It was Sturtevant's first academic appointment and his first taste of administrative power.[70] Morgan agreed that Sturtevant, as division head, would have the authority to hire new faculty, decide on teaching and curricula, and generally run his division in any way he saw fit. It was a remarkable and unprecedented delegation of power by one who had been long accustomed to decide everything himself. Everyone was overjoyed by Sturtevant's new status but thought it was too good to be true: "We all feel," Schrader wrote, "that the Boss will be totally incapable of keeping his hands off this division in particular [genetics], and that Sturt won't be as independent as we would like to see him."[71] There were early warning signals. For example, Morgan would not spell out to Sturtevant exactly what he meant by "practical independence in running your division," and he kept his negotiations with Caltech secret (characteristically) while leaving Sturtevant to take care of the messy practical details of a transcontinental move. Schrader thought it was the first time that Sturtevant, Morgan's confidante and understudy, did not know what was going on behind the scenes.[72]

What Sturtevant's friends feared soon came to pass. Morgan did in fact prove incapable of keeping his hands off the genetics division. Indeed, he ended up making all the important decisions himself, including the key decisions about appointments, about which Morgan and Sturtevant had quite different ideas. Although it must have been a bitter pill, Sturtevant simply deferred to the Boss's authority, as he always had, without protesting or fighting back. "As Sturt says with his shrug of the shoulder," Schrader reported, 'you know whose policies will be followed and whose men chosen.' "[73] Before long offers were indeed made to Morgan's choices and were accepted, except for Curt Stern, whom Morgan offered only Sturtevant's old job on the Carnegie grant. Sturtevant knew full well Stern would not accept that and would have

70. F. Schrader to C. Stern, 29 Oct. 1927, CS; and Morgan to Merriam, 11 Feb. 1928, CIW.

71. Schrader to Stern, 29 Oct. 1927, CS.

72. Ibid.; and Schrader to Stern, 21 Jan. 1928, CS.

73. Schrader to Stern, 21 Jan. 1928, CS.

offered him a faculty appointment, but he was as usual overruled by the Boss.[74]

Sturtevant, though clearly distressed by Morgan's failure to make good his promise of independence, never asserted himself. Open rebellion was not congruent with his role as Morgan's student and understudy. The Boss was the boss, and Sturtevant deferred to him, retreating into gloomy silence.[75] He would never talk about this episode, even with sympathetic friends like Franz Schrader, and his silence is mute testimony to his discomfort. Sturtevant was both ambitious and loyal. He desired independent authority, knew he deserved it, and clearly resented having it slip away after he had waited for it so long. If personal ambition seemed like disloyalty, as it could well have in that situation, it would indeed have been a most uncomfortable double bind.

As with Muller, Sturtevant and Bridges's resentment was not personal but structural and generational; it was inherent in the hybrid character of the core Carnegie group and in the way it had gradually evolved from a student project. Sturtevant and Bridges occupied positions that offered unparalleled opportunities for production, but they were also cut off from opportunities for career advancement that academics would have taken for granted. Lacking academic appointments, Sturtevant and Bridges could not rise in status through promotion, nor could they as Carnegie "associates," a ladder with one rung. In 1928 Sturtevant confided to Schrader that Stern had done the right thing in declining Morgan's offer of his old job because, unlike an academic job it would offer no chance for advancement.[76] He spoke from experience. The very lack of social hierarchy, which so greatly benefited the group's collective work, complicated the moral economy as Morgan's boys entered illustrious midcareers while remaining dependent on the Boss. They were locked in a situation that preserved a moral relationship that they had long outgrown. How could their relation to the Boss be anything but ambivalent?

Morgan seems to have understood the situation rationally, but he was unable to bring himself to push his boys from the nest. He knew

74. Schrader to Stern, 7 June, 1928, CS; Anderson to Demerec, 16 May 1928, MD; Morgan to Merriam, 26 Apr., 17 Oct., 14 Dec. 1928, CS; Morgan to Stern, 18 May, 3 July 1928, CS; Stern to Morgan, 11 June 1928, CS; Bridges to Stern, 24 May 1928, CS; and Morgan to Merriam, 14 Dec. 1928, 14 Jan. 1929, CIW. See also Allen, *Morgan*, pp. 350–52.

75. Schrader to Stern, 7 June 1928, CS.

76. Schrader to Stern, 30 June 1928, CS.

that Sturtevant and Bridges would be better paid and more independent if they left him and took academic jobs.[77] Yet when offers did come, Morgan moved smartly to keep his boys from leaving, as when Metz tried to lure Sturtevant away. Why would they want to leave, Morgan wanted to know, when they had always been given complete freedom in their work?[78] In 1920 Sturtevant seriously considered taking a post at the University of Texas but, as Muller put it, "was finally persuaded not to try for it," presumably by Morgan.[79] Sturtevant continued to receive attractive offers—three in 1926–28—but each time was persuaded to stay.[80] It is hard not to suppose that Morgan was counting, consciously or not, on Sturtevant's habit of loyalty to himself and the group.

Sturtevant and Bridges enjoyed the benefits of being at the center of *Drosophila* production, with its superb collection of stocks, its concentration of talent, and its ready access to everything that was going on everywhere in the drosophilists' world. But they paid a price for these advantages in their continuing dependence on Morgan, who deferred to them completely in the *Drosophila* work but in financial and political matters could not bring himself to delegate an iota of power. Despite their international status, at home Sturtevant and Bridges remained the Boss's boys. It was not true, as Dobzhansky once stated, that Bridges and Sturtevant were "merely assistants in the Carnegie Institution."[81] But the fact that he could have so caricatured them suggests the ambiguities of identity and status that they had to put up with.

The Fly Group: Caltech Years

The fly group was never more innovative and productive than it was in the 1930s. The new laboratory was designed to continue the easy mixing and collaboration of the Columbia years. Doors between laboratory rooms were always kept open, there were no individual offices or telephones, and workplaces were communal as before. Morgan was in and

77. Morgan to Merriam, 26 Sept. 1923, CIW; see Morgan to Woodward, 2 Oct. 1917, CIW.

78. Morgan to Metz, 26 Feb. 1919, THM-APS.

79. Muller to Huxley, 22 Nov. 1920, HJM-A. Morgan indicated that Sturtevant had been offered the job and declined; he raised Sturtevant's salary to match what Texas offered: Morgan to Woodward, 3 Oct. 1920, CIW.

80. Morgan to Merriam, 14 Dec. 1928, CIW.

81. Dobzhansky, oral history, p. 270.

out (he worked a good deal at Caltech's marine station at Balboa in Southern California), and Sturtevant presided over a group that included a growing stream of visitors from abroad.[82] At the same time, however, the fly group's internal tensions were nearer to the surface and more openly expressed than they had ever been.

Translocation was a wrenching experience for everyone, but especially for the younger members of the group. Almost no one liked Pasadena's hot dry climate, or its small-town provincial isolation.[83] Sturtevant unburdened himself to Otto Mohr:

> The first winter here we were terribly homesick for New York, and felt as though nothing about Pasadena could be worse than it was. The second winter wasn't quite so bad, and this one is a little better still: but we are still far from enthusiastic about the climate, the scenery, or the inhabitants. I doubt if it will ever be possible to develop a real intellectual center in this climate and starting from such a general population of religious cranks and retired farmers and bankers who have come out here to die. I hate to think of bringing up the children in such a place. We have built a house at Woods Hole, and are firmly decided to go there regularly in the summer, as an antidote.[84]

Sturtevant also disliked the desert; to him its dry barrenness made pleasure trips unenjoyable and insect collecting impossible.[85] He did not even buy a house until 1935, and even then he hinted to Demerec that he would seriously consider any reasonable offer of a job on the East coast.[86]

Building laboratories and departments from scratch was also disruptive, since it meant postponing research for years and waging frequent battles with Morgan for the money to equip empty labs. Ernest Anderson, who was in charge of building a laboratory and greenhouses for plant genetics found the task "pretty trying most of the time," as he discreetly put it. Sturtevant was irritated by the constant meddling of

82. Beadle, "Sturtevant," p. 170; and Sturtevant to Mohr, 11 Feb. 1931, OM. There were fewer graduate students than at Columbia at first, but more visitors.

83. E. Anderson to Demerec, 24 Jan. 1931, MD; Sturtevant to Stern, 6 Sept. [1934], CS; and Beadle to Demerec, 15 Feb. 1936, MD. Schultz and Helen Redfield were the exceptions: Schultz to Stern, 12 Feb. 1930, CS.

84. Sturtevant to Mohr, 11 Feb. 1931, OM.

85. Sturtevant to Stern, 12 Apr. 1929, CS; and Schrader to Stern, 11 Feb. 1929, CS.

86. Demerec to Schrader, 17 July 1935, MD; and Demerec to E. M. East, 20 July 1935, MD.

Caltech's chairman, Arthur Flemming, in the planning of the new laboratory, and by Morgan's tolerance of Flemming's interference.[87] Schrader thought that Sturtevant, too, lacked *Rückgrat*—backbone. The fact is, a career spent as a research associate under Morgan had ill-prepared him to exercise authority, and passivity and indecisiveness had become habits. Although Sturtevant was Morgan's heir apparent, Morgan worried about his chronic inability to make up his mind. Sturtevant also complained of a heavy teaching load; Schrader though that Morgan "must have shoved some undergraduate teaching off on Sturt, which he was not supposed to do."[88]

The greatest source of tension was Morgan's habit of controlling everything himself and his aversion to spending money on research equipment. Sturtevant hinted discreetly about that to Mohr: "He [Morgan] is built so that he is unable to let any of us take some of it off his shoulders, even in small matters: yet he hates to do it, because it takes practically all of his time. It troubles me a lot, but I can see no solution."[89] Franz Schrader was more pointed:

Sturt is quite evidently not satisfied out there and he is not the only one of that state of mind. But none of them will talk much about it. . . . So far as I can gather from indirect sayings and rumors, the main trouble lies in the fact that Morgan is the best of research men himself but he has no idea of how other research men should be treated. And this in turn is mixed up with his great unwillingness to spend any money for research. A lot of funny stories are beginning to spread among zoologists about him, as for instance the building of greenhouses without glass (because glass costs too much) etc. etc.[90]

Glassless greenhouses? No wonder Anderson was disgruntled! Dobzhansky also suffered from Morgan's parsimony and resented Morgan's diversion of money from genetics to other departments. Bridges too seemed resentful, and with some reason, since he had to pay for a tech-

87. Anderson to Demerec, 24 Jan. 1931, MD; and Sturtevant to Mohr, 11 Feb. 1931, OM.

88. Schrader to Stern, 11 Feb. 1929, CS; and Morgan to Millikan, [?] Apr. 1939, RAM 18.

89. Sturtevant to Mohr, 11 Feb. 1931, OM. Morgan, he thought, had aged visibly since the move west.

90. Schrader to Stern, 14 Jan. 1931, CS.

nician out of his own pocket to help him with routine stockkeeping.[91] The reconstitution of the Morgan group made Morgan's bad administrative habits more costly and brought Morgan into more direct competition with his young lieutenants in matters of finance and program, which he had been accustomed to control himself. At Caltech the fly group became less like a research institute and more like an academic department, and patterns of authority and deference that had shaped life and work at Columbia no longer worked.[92] It is indicative of the changing nature of the group that the most innovative of the new generation of drosophilists—George Beadle, Dobzhansky, and Schultz— all left the group for independent careers, and only Beadle's departure was not messy.

Last to Go: Jack Schultz

Jack Schultz's career in the fly group is especially revealing of its changing moral economy. Of all the second-generation drosophilists, Schultz was most like the first. He came from a large family of Russian Jews who had fled to New York in the late 1890s to escape reprisals for their socialist activities. Bright, precocious, and independent, Jack loved books and concerts and was drawn into the fly group when, as an undergraduate, he came to work there as a bottle washer to support these habits. By the time he got his Ph.D. in 1929, Schultz had staked out his distinctive role as the fly group's idea man, gadfly, and talker. He was interested in everything new and, Dobzhansky recalled, "spent most of his time going from room to room discussing everything with anybody who would listen." Seminars, to Schultz, were "an anticipated fete or celebration." He was quick-minded and excitable and had a tendency to stutter.[93] He exemplified the culture of the *Drosophila* elite.

In some ways, however, Schultz was different from the older dro-

91. Sturtevant to Mohr, 11 Feb. 1931, OM; Viola Curry to Dunn, 13 Oct. 1933, LCD; Morgan to Merriam, 6 Nov. 1935, CIW; and Dobzhansky to Demerec, 7 Jan. 1934, MD.

92. CIW president John C. Merriam worried that the Carnegie team might lose its distinctive identity. Officially it did not, but in practice it did: Merriam to Morgan, 23 Oct. 1928; Morgan to Merriam, 5 Nov., 14 Dec. 1928; all in CIW.

93. T. F. Anderson, "Jack Schultz," *Biog. Mem. Nat. Acad. Sci.* 47 (1975): 393–427, on pp. 393–96, 409; Dobzhansky, "Answers to W. Provine's questions," nd [Apr. 1975], TD file Mayr. Anderson to Stern, 16 Feb 1973, CS.

Figure 4.9 Jack Schultz as
a graduate student at Co-
lumbia, circa 1925. Cour-
tesy of Jill Schultz Frisch.

sophilists. He took more pleasure in beginning projects than finishing
them. His plans were, as Dobzhansky put it, "always subject to rapid
changes." He hated writing and would do almost anything to avoid it.
Even when he finally realized, in the late 1930s, that his career depended on getting his work published, no amount of pushing and bul-
lying by friends like Dobzhansky could get him to do it. "He simply will
never publish his work," Dobzhansky concluded sadly. "Well, it is too
bad, but who and how [*sic*] can help him!"[94] As a result of his bad
work habits, Schultz published a good deal less than other members of
the group.

His paper trail is not, however, a good indicator of his contributions to the intellectual life of the group. No one did more than Schultz

94. Dobzhansky to Demerec, 10 Mar. 1939, 7 May 1940, MD; see also ibid.,
16 Nov. 1939, 10 Jan., 26 Mar. 1940; Dobzhansky to Stern, 17 Feb. 1940, CS;
and Anderson to Stern, 16 Feb 1973, CS.

to extend the range of classical genetics into the biochemistry and bio-physics of genes and gene action. He deeply influenced both Beadle and Dobzhansky, as we shall see, and was continually engaged in collab-orative projects. Perhaps because he found it hard to write up his own work, he was always ready to help others in refining a point or resolving a knotty problem of interpretation, as Lillian Morgan wistfully recalled after Schultz had departed.[95] In that regard Schultz resembled Sturte-vant, though he was intellectually bolder and more volatile. As Sturte-vant's successor on the Carnegie team, he was unfettered by any formal obligations to the Carnegie stocks, and so could act like a freewheeling research professor, following his genius wherever it led.

Schultz's situation changed in the late 1930s, however, when Bridges's death left him with the responsibility for preserving the Car-negie stocks and the Carnegie grant. Morgan had always worried that the Carnegie Institution would take his retirement as an excuse to ter-minate the grant, and he was right. When Bridges died in 1938, Van-nevar Bush, the CIW's tough-minded new president, refused to replace him, making ominous remarks about "diminishing returns" from the *Drosophila* project.[96] Morgan continued to believe, despite growing evi-dence to the contrary, that the Carnegie stock collection would be the key to the continued survival of the group. The trouble was that no one in the fly group was using the *melanogaster* stocks in his research. So Morgan concocted a plan for a big mapping project that could only be done with the Carnegie stocks, and since Schultz was already on the Carnegie payroll, he seemed to Morgan the one to shoulder the bur-den. "Since there is no one in the department using the melanogaster stocks," he wrote, "I feel it inappropriate to ask the Carnegie Institute [*sic*] or the California Institute to support the work unless this stock was utilized for the purpose for which it has been built up, and as you are the only person available who is competent to carry on this line of work it will be necessary for you to turn your attention to that if our program is to continue."[97]

Schultz was not pleased with the prospect of being Bridges's succes-sor. He was used to being a freewheeling intellectual, not an employee with chores to perform. Besides, he was no longer interested in main-stream genetics. He had just spent two years at Stockholm with the biochemist Torbjörn Caspersson, studying the biochemistry and bio-

95. Tyler to Schultz, 14 Jan. 1946, JS.

96. Morgan to Bush, 25 Jan., 28 Sept. 1939; Bush to Morgan, 30 Jan., 10 Apr., 19 Sept., 4 Oct. 1939; all in CIW.

97. Morgan to Schultz, 9 Oct. 1939, JS.

physics of chromatin, and he planned to continue his work on gene chemistry at Pasadena. Morgan, however, expected Schultz to defer to his authority and to the group's need, as Sturtevant and Bridges always had. When Schultz was late returning from Europe, Morgan lectured him sharply on his duties, demanded that he return at once, and stopped his salary until he did. Schultz had to borrow $300 from Dobzhansky to get home.[98]

Feelings of deference ingrained in the first generation of Morgan's boys were much attenuated in the second. Schultz's reply to Morgan's demand was cool, remarkably cool considering Morgan's peremptory tone. "My own feeling," he confided to Sturtevant, "is that one can't pick a fellow to continue taking care of the stocks."[99] He hoped that Morgan might approve a project that would fit his new interest (e.g., mapping chromosomes spectroscopically and comparing them to salivary maps), but Morgan refused to pay for the expensive spectroscopic apparatus that he would need. In the end Schultz did concoct a scheme that satisfied Morgan with minimal cost to himself. As Dobzhansky irreverently put it, he "cooked up a plan of a work to be done by Mrs Curry [the stockkeeper], the work to be such as to require all the available stocks and hence justifying the existence thereof."[100] But Schultz resented the imposition, neglected his official responsibilities, and took every opportunity to escape to places that offered him more scope for his own work. Sturtevant made disparaging remarks, and there were tiffs with Morgan and sharp little letters chiding him brusquely about his obligations to the Carnegie group.[101]

Morgan's somewhat arbitrary treatment of Schultz should not be misunderstood. He had a warm spot in his heart for his bright and charming if sometimes errant protégé, the last Ph.D. of the Columbia era, his Benjamin. But Morgan expected him, as one of his boys and as an employee of the Carnegie Institution, to serve the group.[102] The trouble was that with Bridges dead and Beadle and Dobzhansky gone,

98. Dobzhansky to Demerec, 27 Oct. 1939, MD.

99. Schultz to Morgan, n.d. [Oct. 1939], JS; and Schultz to Sturtevant, n.d. [1939], JS (quotation).

100. Dobzhansky to Demerec, 8 Nov. 1939, MD; and Morgan, "Suggested program of future work," to Bush, 19 Oct. 1940, CIW.

101. Dobzhansky to Demerec, 23 Nov. 1939, MD; and Schultz to Caspersson, 31 Aug. 1940; Morgan to Schultz, 8 July 1941; and S. Emerson to Morgan, 4 July 1941; all in JS.

102. Tyler to Schultz, 14 Jan. 1946, JS; Schultz to Morgan, 3 Jan. 1929, JS; and Morgan to Merriam, 17 Oct. 1928, CIW.

the fly group as Schultz had known it no longer existed. ("Poor Pasadena," Richard Goldschmidt intoned, "it will be completely blown up and sink into oblivion.")[103] Changes in the research front left the Carnegie stocks more a burden than a valuable tool. Morgan's retirement and Bush's termination of the CIW's grant in 1941 were anticlimactic. In 1942 Schultz accepted an offer to start a genetics research unit at the Lankenau Hospital in Philadelphia. He would have stayed at Caltech, but Sturtevant, ever cautious, dared only offer an assistant professorship, and Morgan declined to interfere, though he later felt it had been a mistake to let Schultz go.[104] With Schultz's departure the seasons of the fly group's life cycle came full circle.

Conclusion

We have focused in this chapter on the social technology of the fly group and on its moral economy. We have seen how the drosophilists organized experimental work to exploit *Drosophila*'s remarkable capacity to produce material for new experiments, and how they managed their social and moral relationships to ensure access to a communal instrument and limit individuals' claims to credit for collective achievements. The society of the fly group was not consciously constructed to do these things; rather, it evolved out of a project designed by and for a couple of precocious graduate students, and it drifted into permanence through Carnegie funding. The result of this evolution was an institution that was a hybrid of research institute and academic department, and a moral economy that perpetuated habits of deference and authority characteristic of the early stages in the cycle of scientific careers. Morgan remained the Boss in financial and political matters, while "the boys" became his equal in experimental skill and output. This disparity between authority and status engendered generational tensions, which erupted in periods of stress.

103. Goldschmidt to Dobzhansky, 19 Feb. 1940, TD.

104. Schultz to Sturtevant, n.d. [Apr. 1942]; Sturtevant to Schultz, 22 July, 20 Oct. 1942; Schultz to Stadler, 26 Aug. 1942, n.d. [Aug. 1942]; Schultz to Morgan, 22 Aug. 1942; Schultz to Goldschmidt, 9 Dec. 1942, 9 Dec. 1943; Goldschmidt to Schultz, 17 Dec. 1942; Tyler to Schultz, 14 Jan. 1946; all in JS; and Schultz to F. S. Hammett, 17 Apr. 1942, and Schultz to S. P. Reimann, 29 Aug., 23 Oct. 1942, all in JS file Institute for Cancer Research. On Sturtevant as leader see Morgan to Bush, 28 Sept. 1939, CIW; and S. Emerson to Beadle, 21 May, 20 June 1945, GWB 1.25.

Although the culture of the fly group owed a good deal to the personalities of its leading members, it was mainly shaped by forces that structure life and work in all small groups: the tensions of generational succession; the imperatives of participation, equity, and authority; the pragmatic need simultaneously to produce and change. The original fly group was unlike any of its subsequent competitors, but it was not sui generis, and its history offers general insights into the practices of experimental scientists.

Questions of participation, equity, and authority also arise in connection with the relations between the fly group and other centers of *Drosophila* production. For almost thirty years the fly group was the center of an expanding international network of drosophilists, its news exchange, its source of trick mutants and craft knowledge, its exemplar of the drosophilists' experimental culture. The fly group's customs and working habits derived not just from the fly itself and the fly group's internal dynamics, but also from the group's external relations with other producers. Only a network of users could fully exploit the productive potential of the breeder reactor. And only wide participation in its practices could give these practices a cosmopolitan presence. The experimental culture of the fly people first evolved in a special local ecology, a cultural island, but it became cosmopolitan as it spread around the world via a vigorous commerce in mutant stocks and the tricks of the trade that went with them. That system of exchange is our next topic.

FIVE

The *Drosophila* Exchange Network

I T WAS THE CUSTOM AMONG DROSOPHILISTS from the earliest years to share mutants and know-how freely with each other. This custom was not just an occasional one: a good proportion of the drosophilists' correspondence deals with swapping stocks. And it was not just worked-out stocks that were shared: even those that were actively producing were regarded as communal property. Drosophilists were not alone in having such an exchange system. Maize geneticists had a similar system quite early, and bacterial and phage geneticists later made it a standard practice of molecular biologists. Few experimental biologists, however, engaged in exchange as systematically and as earnestly as drosophilists. Given the pervasiveness of biologists' systems of exchange, it is surprising that there has been no systematic study of how these systems originated and how they work.

There can be little doubt that the fly people's custom of exchange was a significant cause of the growth and spread of their novel experimental practice. Ready access to the means of production made it easy for newcomers to enter the new field, and harder for skeptics to doubt the credibility and value of drosophilists' new conception of genetics. Exchange of standard stocks and standard practices helped to integrate geographically dispersed practitioners into a working community. It made a standard organism possible, since "standard" is by definition what everyone does. Exchange accelerated the pace of production by widening participation and made it possible to exploit fully the productive potential of the breeder reactor. Formal scientific publications spread the *word*, but personal exchange of working tools was probably more effective than publications, with all their artful elisions of craft methods, in spreading the *work* of the fly people.

The custom of free exchange was fundamental not only to the drosophilists' productive economy but also to their moral economy. Drosophilists were unabashedly proud of their tribal custom, regarding it as an "unwritten law," a badge of citizenship, a mark of professional identity. Sharing tools was a sign of membership in a special community, a guarantee of participation with all its rights and responsibilities.

This mixture of practical and moral imperatives is apparent in a letter from Morgan to his patron, Robert S. Woodward, in 1917:

> We make a point of supplying any individual or group of individuals with any material in stock, not only material that has been studied by ourselves but also material as yet unpublished if it can be utilized. The method of locking up your stuff until you have published about it, or of keeping secret your ideas and progress has never appealed to me personally, and I think as a simple matter of policy that such a procedure is as injurious to the student as it is to the progress of science, which we profess to have most at heart. It may be that we can claim no special virtue here, for Drosophila is like the air we breathe—there is enough for all.[1]

Woodward agreed, adding that the world of manufacturing and commerce would benefit no less if the principle of community ownership were also applied to technological inventions.[2] In 1933 Milislav Demerec and Calvin Bridges asserted that the accomplishments of *Drosophila* genetics were largely due to "the broadmindedness of the original Drosophila workers who established the policy of a free exchange of material and information."[3] There had been no conspicuous abuse of the system, they claimed, in over twenty years.

The moral economy of free exchange implied a conception of scientific discovery and reward that was as much collective as individual. As Demerec observed, in a closely knit group of people who worked with the same material on the same basic problems, it was largely a matter of chance who made a discovery. If one person did not, some one else would soon enough. The whole idea of discovery as a discrete event for which personal claims could be staked seemed a little arbitrary and artificial. The less the personal element was allowed to intrude, Demerec thought, the better it was for everyone.[4]

This vision of sharing and mutual aid is, of course, a founding myth of the sort that often accompanies the creation of new modes of experimental life and new disciplines.[5] It was a normative moral code and

1. T. H. Morgan to R. S. Woodward, 25 July 1917, CIW.

2. Woodward to Morgan, 28 July 1917, CIW.

3. C. B. Bridges and M. Demerec to *Drosophila* geneticists, 10 Nov. 1933, HJM-C; cf. Demerec, foreword, *DIS* 11 (1939): 6.

4. Demerec, foreword, *DIS* 2 (1934): 3–4.

5. Simon Schaffer, "Scientific discoveries and the end of natural philosophy," *Soc. Stud. Sci.* 16 (1986): 387–420.

cannot be taken uncritically as descriptive of real behavior. Morgan and Demerec articulated their ideals in formal settings for practical reasons: Morgan, to justify his activities to his patron, and Demerec, to win loyalty to his newly inaugurated newsletter, the *Drosophila Information Service*. Their utterances were performances that were meant to influence others. But if selflessness and altruistic sharing cannot be taken as descriptive of real behavior, neither can they be dismissed as mere rhetoric. Similar sentiments appear in private letters, as, for example, when Morgan wrote Fernandus Payne not to refrain from working on any *Drosophila* species for fear of treading on the toes of the Columbia workers: "We have no copyright on any of them. . . . Therefore, don't let what we are doing influence you against working on any or all of these bugs."[6] Nor should we cynically conclude that normative precepts do not in fact guide real behavior. Obviously they do: the literature on moral economy leaves no doubt of that.

In fact, instances of really damaging fights among drosophilists are remarkably rare, although personal rivalries were common and sometimes bitter. The moral economy of exchange did not suppress individualistic or competitive feelings; but it did keep the public expression of these impulses within bounds. Drosophilists were a remarkably civil and uncontentious lot, far more than were, for example, general physiologists, biochemists, or endocrinologists, and their civility clearly was an integral part of their distinctive mode of production and exchange.

We need, then, to discover how drosophilists' system of exchange served as an instrument of group production and as a moral code for regulating competitive feelings and privileges. Who gets access and who gets credit? Who sets agendas and who emulates them? What privileges attach to working in a large center of production and what responsibilities to less well placed producers? These were fundamental issues for drosophilists, as they are for any community, and the *Drosophila* exchange network was perhaps the most important everyday activity in which such issues arose and were addressed and managed, if not resolved. How did the exchange system arise, and how did it work? What were the limits beyond which sharing and selflessness were regarded as illegitimate intrusions on individual rights? How was it experienced by those who participated in it? What were its long-term effects on practice?

6. Morgan to F. Payne, 28 Nov. 1915, FP.

The *Drosophila* Exchange Network

Exactly where the custom of exchanging stocks came from is not entirely clear, but there are several places where it was deeply rooted in
biologists' working culture. Exchanging specimens had long been the
custom among natural historians, especially among entomologists. The
natural variety and abundance of bugs, and the ease of collecting vast
numbers of them, encouraged a habit of active barter. A similar spirit
of self-help also prevailed among the experimental morphologists who
flocked to the many marine biological stations created around the turn
of the century on the model of the celebrated Naples Station in Italy.
This spirit of collective self-help was especially well developed in the
Marine Biology Lab at Woods Hole, Massachusetts. Its summer sessions
were designed to give underprivileged and overworked teachers of
general biology in small colleges access to resources for research and
to bring them into contact with vanguard researchers. For rank-and-
file biologists, summering at Woods Hole was a way of participating,
even if marginally, in scientific high culture. For the elite, it was a way
of recruiting talents, increasing production, and building constituencies for vanguard research lines. It was a public service and a moral
duty, as well as being fun. Wide participation in research by rank and
file as well as elite was highly valued for both practical and moral reasons.[7] Morgan was a mainstay of this system, leading the entire fly
group with their families, stocks, and paraphernalia each summer from
New York to the shore of Buzzards Bay.

The drosophilists' custom of exchange and participation was thus
an amplified form of habits that were generally prevalent in their professional circles. The amplifier was *Drosophila*, the breeder reactor; it
removed any possibility that access to the means of experimental production might seem a zero-sum game. The fly's fecundity made the
cultural values of sharing and self-help robust and resistant to the normal temptations of selfish interest. There was no advantage to be

7. Philip Pauly, "Summer resort and scientific discipline: Woods Hole and
the structure of American biology, 1882–1925," in Ronald Rainger, Keith R.
Benson, Jane Maienschein, eds., *The American Development of Biology* (Philadelphia: University of Pennsylvania Press, 1988), pp. 121–50; Jane Maienschein,
ed., introduction to *Defining Biology: Lectures from the 1890s* (Cambridge: Harvard University Press, 1986); and Maienschein et al., "The Naples Zoological
Station and the Marine Biological Laboratory: One hundred years of biology,"
Biol. Bull. 168 suppl (1985): 1–207.

gained in restricting access. Indeed, the overabundance of productive material made sharing not just a virtue but a necessity for a group that produced far more material than they had the resources to work up. The values and habits of sharing and participation grew out of the concrete realities of Drosophilists' daily practices.

We may catch a glimpse of the exchange system taking shape in 1911–12, when Morgan was desperately trying to keep up with the flood of mutants. Unable to cope by himself, he turned to teachers of biology in small colleges, persuading them to take charge of some aspect of the *Drosophila* work and eventually count it toward a Ph.D. dissertation. Morgan provided the necessary mutant stocks, and his recruits worked on them as they could between classes and in summers at Woods Hole. John Dexter was recruited in this way to do the initial analysis of *beaded* while teaching at Northland College (he finished his Ph.D. at Columbia in 1912–14). Clara Lynch worked on her flies for seven years while at Smith College, and Shelley Safir for eight years while teaching high school biology in New York City. By sharing tools, Morgan got the *Drosophila* work done, and his scattered recruits bootstrapped their way to Ph.D.s.

In farming out drosophilas to small peripheral producers, Morgan was making use of those underemployed as an accidental by-product of the overexpansion of collegiate education in late-nineteenth-century America. In the 1900s and 1910s a generation of teachers with B.A. or M.A. degrees had to deal with a credentialing inflation, as reformers made the Ph.D. the standard credential for college teaching.[8] A generation of experienced, often talented, and underemployed biologists was thus available for doable research projects, and they needed only to be provided with tools and access to vanguard problems. Morgan simply used this informal system of nonresident graduate study to expand the *Drosophila* project fast and cheaply. Probably he got to know Lynch and Dexter at Woods Hole. In any case, their arrangement was inexpensive, low-risk, and mutually beneficial. Morgan got relief from the flood of mutants, and his recruits got degrees at a pace and a price that they could manage. In the careers of early drosophilists like Dexter, Lynch, and Safir we can see how the *Drosophila* exchange network may have evolved out of biologists' traditional customs of participation and self-help, under pressure from the breeder reactor.

8. Robert E. Kohler, "The Ph.D. machine: Building on the collegiate base," *Isis* 81 (1991): 638–62.

Centers and Peripheries

The moral economy of the exchange network was also a product of
the peculiar distribution and stratification of drosophilists, with one
predominant producer at the top of a system of many small producers.
Because it was not threatened by serious competition, Morgan's fly
group could afford to be generous in distributing the tools of production to potential competitors. A marketplace of production that was
broadly based and highly concentrated could afford the values of sharing and noblesse oblige. A system of more equal producers or of limited resources would doubtless have encouraged customs that were
more individualistic and restrictive, as was the case in other areas of
experimental biology.

It is quite amazing how long the Morgan group remained the only
large *Drosophila* group (see Appendix). Many individuals worked occasionally on *Drosophila,* but for two decades Morgan's was virtually the
only group that could sustain a comprehensive program of research.
Only in the late 1920s and early 1930s did new centers of production
arise—Cold Spring Harbor, two Kaiser Wilhelm Institutes at Berlin, the
University of Texas—that could challenge Morgan's group in mainstream lines of work. Not until the late 1930s did all major universities
feel incomplete without a contingent of drosophilists. The original fly
group was able to control the pace and agenda of *Drosophila* genetics
for nearly a whole generation.

The intellectual predominence of the fly group was probably even
greater than its share of total output would suggest. In the early 1920s,
when the group was producing less than a tenth of all papers in the
field, it produced many of the most novel and productive new experimental systems: the attached-X and triploid mutants, deficiencies, mosaics, and so on. This remarkable control of the production of new
tools was undoubtedly a consequence of the fly group's strategy of foraging widely in many lines and pursuing only the most novel and productive for publication. The fly group members were virtuosos, in
short, in the modern art of skimming off the cream, and their ability
to stay continually in the vanguard was a formidable barrier to would-
be competitors. Yes, the group had exceptionally talented people and
an enviable level of material support, but the crucial element was the
way they deployed these endowments. It was the fast pace of innovation
that made it difficult for small producers to compete on a broad front.

Most small producers, as a result, pursued the strategy of specializ-
ing in some small line in which they might compete with their more
privileged colleagues at the center. At Cold Spring Harbor, Charles

Metz concentrated on comparative genetics and interspecies hybrids (more on that later). At the University of Illinois, Charles Zeleny built a colony of drosophilists around the physiology and development of the bar-eye mutants. At Amherst College, Harold Plough concentrated on temperature effects in development. Muller's decade-long project on natural mutation rate is another example of this strategy, as is the embryological work of Ruth Howland and her circle at New York University. N. W. Timofeeff-Ressovsky began with a broad program of *Drosophila* studies at the Kaiser Wilhelm Institute for Brain Research, but from 1928 on concentrated almost exclusively on radiation and mutation. At the University of Texas, Muller, John T. Patterson, and Theophilus Painter systematically exploited their new techniques of X-ray mutation and salivary chromosome cytology. Demerec at Cold Spring Harbor, Otto Mohr at Oslo, and Curt Stern at Berlin worked in the fly group style, but got the resources to form large groups only in the 1930s, if then: Mohr, stuck in a department of anatomy, had to wait even longer.

Specialization was an effective strategy for small producers because it avoided the risks of competing directly with the fly group in lines they had invented. Specialization had its own risks, of course: if a line ran dry or ran aground on an experimental artifact, one could be left empty-handed. Metz, Zeleny, and Howland all experienced such setbacks, for example, when species failed to hybridize, and reverse mutations in *bar* turned out to be an artifact of unequal crossing-over, and embryological transplantation failed to live up to its promise. Metz dropped *Drosophila* altogether, as we shall see, and Zeleny sometimes yearned to "leave the pesky Drosophila" and return to embryology, his first love.[9] Lack of resources and the hazards of specialization made it difficult for single researchers at the periphery to sustain their operations.

Thus the fly group dominated the production of *Drosophila* genetics for nearly two decades. It was the undisputed center of production and fashion setter in a system of dispersed, small, and intermittent producers who tagged along or pursued bylines that the fly group did not care to pursue. Roy Clausen, in E. B. Babcock's group at Berkeley, would have liked to work on nondisjunction and XXYY females, hot topics in 1920, and wrote Morgan to inquire if they were on the fly group's agenda. They were indeed: Morgan replied that Bridges was well along on nondisjunction and that the XXYY work had been com-

9. C. Zeleny to Ross Harrison, 25 Jan 1921; and Harrison to Zeleny, 31 Jan. 1921; CZ 4.

pleted.[10] No chance for small players there. In 1939 Warren Spencer feared that his situation at Wooster College would make it impossible to keep up in the race to find new species in the wild: "I do hope," he wrote Sturtevant, "that you and Dobzhansky and Patterson and his crowd leave a few odd sub-species or species still to be collected by the time I get time and money to take a crack at it. . . . But I suppose if I'm to find anything new I'll either have to hurry up or take a trip to the Abyssinian Alps or the Mountains of the Moon, either one of which might prove good collecting grounds but rather far from home."[11] Joseph Krafka, frustrated by a heavy teaching load at the University of Georgia, joked ruefully that he would have to publish an account of the problems that he might have worked on if he had the time.[12] He finally improved his lot by giving up zoology and taking a job in embryology in the university's medical school.

The fly group did not have a monopoly, exactly, but they did have the ability to control any significant new lines of research if they chose to do so. The self-confidence that came from knowing that they could prevail in any competition clearly shaped their attitudes about sharing the tools of the trade. It encouraged generosity and sharing between center and rank and file. It also fostered a kind of benevolent paternalism, whether the fly group consciously intended it or not, and this tendency was reinforced because many of the smaller producers (at least in the U.S.) were fly group alumni, who retained habits of loyalty and deference from their apprenticeship experience. (Muller and Edgar Altenburg stand out as exceptions; and some graduates of W. E. Castle's school at Harvard, especially John A. Detlefson, displayed a combative spirit, as did Europeans like Richard Goldschmidt.)[13] A tendency for exchange relations to be benevolent and deferential was thus inherent in the production process and in the way that the community of *Drosophila* geneticists evolved historically from a single source.

The *Drosophila* exchange system was not unlike the preindustrial moral economy of provision, in which landowners took responsibility

10. R. E. Clausen to Morgan, 17 May 1920; and Morgan to Clausen, 28 May 1920; UCDG.

11. W. P. Spencer to Sturtevant, 2 Feb. 1939, AHS.

12. J. Krafka to Zeleny, 14 May 1923, CZ 7.

13. On Detlefsen see Morgan to Zeleny, 15 Feb. [1921], 22 Apr. 1921, CZ 5; on Goldschmidt see Sturtevant to O. Mohr, 4 Mar. 1921, OM. Sturtevant was also peeved at Clarence C. Little, another Harvard scion, who he thought had poached Bridges's work.

for feeding the poor in hard times and enjoyed the privilege of deference and social status in times of abundance. So, too, the fly people took for granted that stocks and know-how would be available from the Morgan group when needed, and they deferred to them as the source of *Drosophila* wisdom and virtue. Sharing in the means of production was held as a golden rule, and individualistic competitive behavior was discouraged at the same time that inequalities were accepted as natural. Elite status was acknowledged in exchange for public service and the right to participate, even if in a small way.

Costs and Benefits of Exchange

The *Drosophila* exchange system depended less on altruism than on enlightened self-interest. Certainly, recipients of the fly group's stocks benefited, but the fly group may have benefited even more, especially in the early years. For example, making *Drosophila* freely available clearly helped establish the new and controversial practice of genetic mapping as the dominant form of experimental heredity. Practices that are limited to one place are easily attacked as merely local, idiosyncratic deviations.[14] Morgan knew well that strength lay in numbers. In 1915, for example, he urged Fernandus Payne not to worry about duplicating results, "for the more people working in that field the better for everybody concerned."[15] Familiarity instilled trust in new methods. Access to tools was no less essential for recruiting graduate students: who would invest in *Drosophila* knowing that it was would be available only in the place where it was invented? Free access also brought inventors of specialized tools visibility, trust, and prestige. In 1929 Demerec was eager to help a young plant geneticist get started with *D. virilis*, for example, because it would help to establish a species associated with him and his laboratory, developed by Metz and himself, as a standard laboratory tool.[16]

Dispersing stocks was also insurance against accidental loss. When the fly group's culture of *notch* died out in 1919, for example, Morgan could hope to retrieve it from one or another of the people to whom

14. Adi Ophir and Steven Shapin, "The place of knowledge: A methodological survey," *Sci. Context* 4 (1991): 3–21.

15. Morgan to Payne, 29 Sept. 1915, FP.

16. Demerec to E. E. Dale, 17 Oct. 1929; and Dale to Demerec, 14 Oct., 7 Dec. 1929; MD.

he had sent stocks in the past.[17] The system did not always work. When the fly group lost their original stock of *truncate*, it transpired that no one had preserved it. Everyone had relied on someone else to maintain a stock, which had been replaced by improved models but which was irreplaceable for redoing old experiments. Muller admitted that he had sacrificed maintenance for current research: "I thought that most of the stocks would be replenishable from Columbia in case of need," he wrote ruefully to Otto Mohr, "but . . . it so happens that Columbia too has lost the truncate stocks which we both depended on."[18] Generally, however, the exchange network was cheap and effective insurance against the loss of valuable tools. Later, when Muller's stocks were wiped out (a window was left open to the summer heat) he had only to ask colleagues for replacements (he was too embarrassed to ask Morgan for a full set).[19]

The custom of exchange also gave the Columbia group rapid access to breaking news from every corner of the drosophilists' world. Every exchange of stocks was an opportunity to share news and talk shop. Bridges's shipments of stocks were often accompanied by technical advice, and his critical appraisals of research plans and manuscripts saved many drosophilists from making fools of themselves in print.[20] In return, Bridges learned of results of work that was done with his stocks well before they were published. Being at the center of an active communication network was one of the greatest benefits of working in the fly group. Knowing what everyone was doing, the fly group could pick up on important or overlooked leads and avoid dead ends. "Write *all* the crazy things," Bridges exhorted Stern, "vastly entertaining and suggestive of new leads."[21] Sometimes swapping mutants led to full-blown collaborations, as happened with Muller and Hans Grüneberg.[22]

Finally, supplying stocks spread the ideas of the new genetics by

17. Morgan to Payne, 11 Jan. 1919, FP. For similar cases see J. Schultz to Demerec, 25 Jan. 1934, MD; Bridges and Sturtevant to Mohr, 15 Sept. 1920, OM; and Bridges to Mohr, 13 Mar. 1923, OM.

18. H. J. Muller to O. Mohr, 8 May 1923, HJM-A; see also Morgan to Mohr, 5 Nov. 1923, THM-APS.

19. Muller to Clausen, 26 Sept. 1923, UCDG.

20. Morgan to E. B. Babcock, 4 Nov. 1918; Morgan to J. L. Collins, 18 Mar. 1918; and Clausen to Morgan, 1 July 1919, 17 May 1920; all in UCDG; and Bridges to L. C. Dunn, 22 Nov. 1937, LCD.

21. Bridges to Stern, 1 Feb 1928, CS.

22. Muller to H. Grüneberg, 31 Oct., 4, 31 Dec. 1936, 11 Feb. 1937; and Grüneberg to Muller, 7 Jan. 1937; all in HJM-C.

giving students hands-on experience with the way they were produced and used. By the late 1910s Bridges and others were routinely sending stocks for classroom demonstrations or student exercises. Patterson began this practice at the University of Texas as early as 1913 or 1914.[23] Zeleny began to supply stocks around 1917 to a network of small colleges and universities in the South and Middle West (e.g., the Universities of Chattanooga, Kansas, Oklahoma, New Mexico, Pittsburgh, and Texas, and Allegheny and Procopious colleges).[24] Supplying stocks for teaching was no small burden. Sometimes Zeleny was asked to supply directions for culturing, breeding, and handling stocks, and colleges had to be continually resupplied, since there were no researchers there to maintain stocks, and cultures died out every summer and had to be "ordered" again each fall.[25] But if supplying teaching stocks was a nuisance, the major centers did it uncomplainingly (if not always promptly) because it was in their interest to have potential recruits and allies witness and participate in their novel mode of experimental work.

Because the Columbia collection was for a long time the only comprehensive source, Bridges and his colleagues bore most of the burden of supply. They did so willingly because they felt morally obligated and because it paid dividends. The exchange system spread the fly group's practices and made them standard practices. It made a local mode of practice trustworthy and legitimate by spreading it among a cosmopolitan community of practitioners. *Drosophila* exchange was the material basis of the fly people's system of recruitment, employment, and communication. It became essential for making careers, staying *au courant*, and participating in the business of experiment.

Rules of the Moral Economy

What, then, were the unspoken rules of etiquette in the moral economy of *Drosophila* exchange? Reciprocity was one: the privilege of receiving stocks entailed the obligation to reciprocate. Exchange was an

23. J. T. Patterson to Zeleny, 18 Mar. 1920, CZ 5.

24. W. S. Adkins to Zeleny, 24 Oct. 1917, 31 Jan. 1918; H. D. Fish to Zeleny, 24 Feb. 1921; R. Green to Zeleny, 24 Jan. 1921; H. S. Jurica to Zeleny, 11 Dec. 1920, A. Richards to Zeleny, 21 Jan. 1921; G. W. Vanzee to Zeleny, 23 Sept. 1922; H. De Bruine to Zeleny, 14 Feb. 1928. Zeleny to Patterson, 21 Jan. 1920; Zeleny to B. O. Severson, 9 Oct. 1918; all in CZ (boxes 3, 4, 5, 8, 9).

25. A. O. Weese to Zeleny, 10 Oct. 1919, 9 June, 23 Sept. 1920, CZ 7.

ongoing, low-level potlatch. Charges were not usually made for stocks, though the Morgan group did ask recipients to return mailing tubes and pay postage.[26] However, it was the custom to pay a dollar or two for stocks used for teaching. Since supplying stocks for teaching was not a reciprocal interaction between active researchers, it was gradually removed from the moral economy into the cash nexus. For example, W. M. Barrows insisted on paying Muller for stocks: "We have been the recipients of so many favors," he explained, "that I would much prefer to buy these outright."[27] When the supply system was formalized in the 1930s, requests for stocks for classroom use were routinely referred to commercial biological supply houses.[28] Nothing better illustrates the moral basis of the reciprocity.

Disclosure was a second rule: recipients of stocks were expected to tell donors what experiments they planned to do and to keep them informed of what the results were, especially if the results came a little too close to the donors' own line of work. Patterson, for example, told Stern what he knew about the structure of translocation stocks that he was sending, and asked Stern to tell him if he turned up anything interesting.[29] When Beadle and Stern swapped suppressor stocks, they also told all they knew about them and how they would be used, just to make sure no one's toes would be trod upon.[30] It was this custom of disclosure that made exchanges of stocks such an effective system of informal communication. Every stock carried with it news of future plans and projects. Exchange also nourished trust and defused suspicions that were entirely natural and healthy among producers who, because they did share tools, were always potential rivals. Failures to disclose were taken as serious reasons to worry about borrowers' intentions.

A third and fundamental rule was that while problems might be owned (temporarily) by individuals, tools could never be: they were the property of the whole community. Well, not quite. The skill and hard work that went into constructing very special tools—such as reagent stocks for mapping, triploids and attached-X mutants, and translocations—did entitle those who did the work to a degree of personal own-

26. Th. Dobzhansky to Sturtevant, 7, 16 Jan. 1934, MD; and Brusston [?] to Stern, 19 Feb. 1940, CS.

27. W. M. Barrows to Muller, 11 Oct. 1926, HJM-C. For a similar case see Weese to Zeleny, 10 Oct. 1919, CZ 7.

28. Demerec to Dunn, 11 Jan. 1934, MD.

29. J. S. Patterson to Stern, 15 May 1933, CS.

30. G. W. Beadle to Stern, 25 Oct., 19 Nov. 1937, 16 Feb. 1940, CS.

ership. It was customary to get permission from the inventors of these valuable and versatile tools before using them, but it was taken for granted that permission would not be refused. The idea was that everyone would benefit from sharing tools that could be applied to more problems than any individual alone could dream up or carry through. It was improper to get mutants from colleagues in order to do faster or better what they had already begun. The idea was to enlarge the usefulness of tools by imagining new things to do with them. On the other hand, it was customary to put no restrictions on what was done with borrowed stocks, so as not to prevent workers from following up on serendipitous leads. Loaned stocks were fair game, and it was up to recipients to know when they might be intruding too far upon donors' own preserve—hence the importance of disclosure and continuing communication. Avoiding unintentional competition required care and tact, for the danger was ever present.

Bridges seems to have regarded even his unpublished work as public knowledge and was usually glad if someone saved him the trouble of writing it up. "Sure, go ahead and say in any publication anything you know I have," he once wrote Stern. "It is always better to have something in [print] than ignored!?" Sturtevant, too, liked to disclaim private ownership: "Use the inversion data . . . in any way you wish," he told Stern. "I'm not planning any general account of the things yet awhile, and of course I don't have any copyright on them."[31] Even Muller did not mind relinquishing results that were produced incidentally to other work (not, however, results that he might someday want to publish himself).[32]

No trade secrets, no monopolies, no poaching, no ambushes— these were the practical rules for establishing trust and working harmony among the fly people.

The rules of the game were social conventions, of course, and those who were not drosophilists born and bred had to learn the rules either by living among drosophilists for a time or by making mistakes and being corrected—a kind of moral apprenticeship. Gert Bonnier made such a mistake, for example, when he used Bridges's high-nondisjunction stock without informing Bridges what he was doing, or telling Otto Mohr, from whom he got the stock. This lapse made his intentions suspect. Stern hesitated when Bonnier later asked him for stocks, and wrote Mohr first to ask if Bonnier could be trusted. Mohr

31. Bridges to Stern, 1 Feb. 1928, CS; and Sturtevant to Stern, 10 Sept. 1927, CS.

32. Muller to Stern, 4 Feb. 1931, 30 June 1930, CS.

assured him that Bonnier's transgression was "due to lack of acquaintance with the scientific practice (coûtume) and not to bad will." A little gentle education was all that was needed:

> Just denote the *special* stocks with the name of their owner, eventually informing him that he may ask for instance Bridges for his permission to use a particular stock if you have any where you may be in doubt. As you know the practice of the fly people has always been to let everybody use any mutants or stocks as *tools* in the attaque of a particular problem. I cannot doubt that Bonnier, who has been at Columbia, now must be fully aware of the "étiquette scientifique." At the time he was just a beginner in scientific work.[33]

It is clear, however, that there were fine points of etiquette that might well be puzzling to outsiders. How, for example, did one know what were "special" tools, to which proprietary rights adhered? At some point, clearly, Bridges's nondisjunction stock was dispersed in so many laboratories that it ceased to be personal and was a free resource to everyone; but when was that point reached? It was as if patents or copyrights had no fixed time limits. Another ambiguous case involved stocks that an individual had constructed or assembled for a special project. They could be reserved for the duration of the project, but when one thing led serendipitously to another, when had one project ended and another begun? In 1933, for example, Patterson declined to send Milislav Demerec some elaborately constructed stocks with multiple deficiencies, because he thought it would be unfair to the graduate student who was working on them. (Besides, Patterson thought, since Demerec only needed deficiencies at known loci, he could easily make his own by X-raying.)[34] When Bridges was in Leningrad without his flies and called on Stern to provide his Russian hosts with stocks, he was careful to say that the list of Stern's "current material" would be for his eyes only and off limits.[35] In the *Drosophila Informa-*

---

33. Mohr to Stern, 15 Aug. 1928, CS. Bonnier wrote Sturtevant in some distress when he learned of his ethical gaffe, but Sturtevant was sure Bonnier had been well meaning and assured Mohr that there were no hard feelings. Sturtevant had been unimpressed by Bonnier's talent for research during the latter's visit to Columbia in 1924. Sturtevant to Mohr, 17 Mar. 1923, 18 Mar., 30 Apr. 1924, OM.

34. Demerec to Patterson, 30 Aug. 1933, MD; and Patterson to Demerec, 22 Sept. 1933, MD.

35. Bridges to Stern, 2 Nov. 1931, CS.

tion Service "personal" stocks were listed separately to make clear that they were not yet freely available.

Limited proprietary rights applied not only to constructed stocks but to found objects: for example, wild drosophilas collected in the field. Since collecting took time and skill, the rules of reciprocity, disclosure, and public ownership applied. In the late 1930s, for example, Sturtevant and Patterson swapped hundreds of flies collected in different regions of the American West, dividing up species that were of particular interest to one or the other. Both parties, however, felt free to exploit unexpected windfalls. Sturtevant told Patterson not to mistake his request for *D. pseudoobscura* as "a 'keepout' sign," but there was no need: Patterson knew he was free to work on anything of interest that turned up in any stocks that Sturtevant sent him, and he expected Sturtevant to do the same with the Texas flies that he sent to Caltech. They agreed that after the two of them had had a crack at the new material, they would then make everything available for general use on the *Drosophila* exchange.[36] It was not always pleasant to be generous, as Maydelle Bishop reported to Muller: "It's like pulling eye teeth to get the flies from him [Patterson], although all he does is to . . . look for sex ratios. It broke his heart to send the pseudo-obscura group to Cal. Tec. He can feel like a martyr indefinitely over that since one stock showed a high female to male ratio."[37]

Conflicts could not always be avoided by generosity and disclosure, however. Ambiguities were inherent in the nature of the tool and in the ways it was used. *Drosophila* problems were so closely related to one another and so likely to unfold in unexpected ways that even with the best intentions two groups could easily end up working on the same problem. The fundamental distinction between tool and problem, so simple in principle, was in practice often hard to make. Trick stocks could be applied to various problems, but they were also objects of study in themselves, and tool could easily and unexpectedly become object. Triploids, for example, could be used as tools to study dosage effects, but the problem of how they were formed by nondisjunction of homologous chromosomes was a problem in its own right. Similarly, mapping a section of chromosome was a routine preliminary to studying particular genes, but if mapping became an end in itself, then it was no longer a tool but an object of study. Drosophilists thus drifted unwittingly into competition. Moral ambiguities could be managed but

36. Patterson to Sturtevant, 27 Sept., 14 Dec. 1938, AHS 4.12.

37. M. Bishop to Muller, 24 Dec. 1938, HJM-C.

Figure 5.1 Milislav
Demerec at his microscope
sorting flies, Cold Spring
Harbor, 1928. Courtesy of
American Philosophical
Society Library, Demerec
Papers, photographs
file 1.

never eliminated, and the rules of exchange were no substitute for personal tact and judgment. Every exchange of stocks was potentially a
delicate probing of knowledge and intentions and an improvised negotiation of rights and interests.

Exchange in Practice

How drosophilists staked claims and avoided conflict can best be seen
in cases where the system failed or nearly failed. Muller's correspondence is an especially good source of such cases: being so hypersensitive about ownership and poaching, he was overscrupulous and explicit
about his understanding of the rules. In 1922, for example, he asked
Zeleny for permission to use his *Bar* mutants in connection with his
project on mutation rate. Although Zeleny was no longer working on
mutation, Muller offered to put Zeleny's name on any paper that came
out of his work: "Since you opened up the field of using bar reverse
mutations as an index of mutation frequency it might justly be regarded as an encroachment for another investigator to step in unasked
and turn your methods to his own profit." If Zeleny wanted to reserve
the problem as his "special field," Muller would understand. Zeleny,

who was far less sensitive to matters of ownership, naturally gave Muller carte blanche.[38] Muller performed a similar little moral ritual with Hans Grüneberg.[39]

Milislav Demerec was also sensitive to proprietary rights. A revealing exchange occurred in 1931 when he asked Sturtevant to lend him his *claret* and *miniature* mosaics for a study of the mutation rate at the *miniature* locus. Sturtevant did not work on mutation rate and was glad to send Demerec the stocks he wanted, but omitted *claret* because he thought it unsuitable for Demerec's experiment. Demerec, assuming that Sturtevant felt some proprietary interest, pressed him for the *claret* mosaics: "I have no desire to take away from you a problem you might be working on," he assured Sturtevant. "I do not expect to publish any detailed data of claret experiments, but, however, would like to refer to general conclusions in so far as they are essential for the analysis of mutable miniature. I would be glad to turn over to you all claret data, or to cooperate with you in any other way you might find desirable."[40] Sturtevant responded smartly to the imputation that he was withholding stocks:

Where did you get the idea that I might be unwilling to have you push the claret and minute-n wing mosaics as far as you wish? Was it because I was so slow in sending the stocks? That was nothing but my chronic laziness. No, I certainly have no slightest objection to your getting all the dope you can—and I should have serious objections if you refrained from publishing it in full detail. I may add that if I did have any such objection, I should be kicked in the part of my body developed by Natural Selection for that purpose, and the objections should be disregarded.[41]

Chagrined, Demerec protested that he had not really meant to impute any such feelings: "I am sorry, I think that I know you well enough not to think what you thought I thought about you."[42] The flustered syntax

38. Muller to Zeleny, 9 Nov. 1922, CZ 7 (see also 5 June 1923, CZ 13); and Zeleny to Muller, 5 Dec. 1922, CZ 13.

39. Muller to H. Grüneberg, 31 Oct. 1936, 4, 31 Dec. 1936, 11 Feb. 1937; and Grüneberg to Muller, 7 Jan. 1937; all in HJM-C.

40. Demerec to Sturtevant, 22 Apr. (quotation), 10 Feb., 7 Mar. 1931, MD. The *claret* mosaics were in *D. simulans,* in which no miniature mutant was known.

41. Sturtevant to Demerec, 27 Apr. 1931, MD.

42. Demerec to Sturtevant, 2 May 1931, MD.

suggests how tangled one could get in trimming competitive feelings to the moral norms of altruism and generosity.

Stern, too, got caught in a web of moral ambiguity in 1928, when he discovered that he and Sturtevant were both doing experiments on dosage effects in triploids containing *stubble* and *bobbed* genes. As he often did, Stern ran his idea past Sturtevant to see if he thought it would work. Sturtevant thought it was a very good idea, so good in fact that Dobzhansky was, at his instigation, already working on it. But he assured Stern that Dobzhansky would do the work as quickly and as well as anyone could.[43] Stern was taken aback. He had hoped that the nature of the gene was a problem that "opened somewhat a field of my own in Drosophila away from the lines you followed." Moreover, he suspected that Sturtevant and Dobzhansky had got the idea from a letter he had written earlier to Franz Schrader. He was aware "that there is no such thing as a copyright in science" but asked nevertheless that Dobzhansky leave the problem to him. (Stern had meanwhile begun work on it, although with stocks that were less ideal for the purpose than Dobzhansky's.)[44]

Stern's was a multiple transgression of the moral order: attempting to divide up intellectual territory, suspecting colleagues of appropriating ideas, and quashing legitimate competition. Sturtevant replied briskly that he had not got the idea from Stern via Schrader, in fact he had been working on it for over a year. Also, and more to the point, Dobzhansky was taking pains to construct just the right stocks so that the results would be elegant and unambiguous. "Now all this may not be good reason for you not to study the case," Sturtevant wrote to his friend, "but it does seem to me to be sufficient reason for Dobzhansky continuing." Blunt words—even too blunt, Sturtevant felt, and he assured Stern in a postscript that he was not at all peeved, just setting out the facts: "When you read the page try to imagine my very best smile and whatever I can do in the way of a pleasant voice. And if you disagree at all with my diagnosis, tell me to go to Hell. For, after all, the affair really is not my business at all."[45] Stern thus received a lesson in the golden rule of the fly people: tools were public property, and problems belonged to those who got to them first and were best equipped to do the job. Doing the work efficiently and well was all that mattered; Sturtevant clearly did not expect Stern to argue with that.

43. Sturtevant to Stern, 1 July 1928, CS.
44. Stern to Sturtevant, 17 July 1928, CS.
45. Sturtevant to Stern, 3 Aug. 1928, CS.

Figure 5.2 Curt Stern sorting flies, Berlin circa 1930. Courtesy of American Philosophical Society Library, Stern Papers, photographs file Stern.

Institutional Rivalries

The rise of new fly groups in the 1930s may have increased the potential for such collisions, straining the moral economy. There were no fewer juicy opportunities, just more drosophilists chasing them. Also, the stakes were higher for larger centers, which were capital intensive and visible, making the costs of failure more widely felt and the pressures to succeed in a big way more intense. Further, more places had the material and human resources to monopolize a major line of work if they chose to, and no track record to reassure others that they, like the original fly group, would play by the rules. Finally, the new generation of drosophilists no longer so exclusively comprised alumni of the original fly group. They had not been schooled in its distinctive moral economy, and it could not be assumed that they would be as inclined as the older generation to restrain competitive feelings.

I do not wish to overstate the change that occurred in the 1930s; open conflicts were still the exception and seem to have engendered no grudges or feuds. But such episodes were more intense, public, and institutional, especially when they involved major centers like Caltech, Texas, and Cold Spring Harbor.

In October 1934, for example, Theophilus Painter was taken aback when Bridges wrote him about a manuscript that he, Painter, had submitted to the journal *Genetics*. Assuming that the leak had occurred through Morgan (an associate editor), Painter fired off an angry letter to the editor, Donald F. Jones:

> As you know, the University of Texas and the California Tech. people are the two main laboratories dealing with Drosophila research in this country. We are in a sense competing, and twice the University of Texas has registered home runs, once the x-ray work and now the salivary gland work. If the California group see manuscripts of our work one year prior to their publication you can well understand that we would be unwilling to submit our publications to your journal. . . . [T]he men in this laboratory are unwilling to allow competitors in our field to enjoy the privilege of examining our work a year prior to publication when we have no opportunity to see theirs."[46]

Painter was further outraged when a newspaper article described a talk that Bridges had given on using the salivary gland chromosomes in cytological mapping, which made it seem that "the affable bushy-head Dr. Bridges" had claimed credit for discovering the salivary gland method and disparaged Painter's work.[47]

Jones passed Painter's letter on to Morgan, who took the opportunity to give Painter (via Jones) a lesson in etiquette: "In regard to the two laboratories 'competing,'" he wrote, "I was under the impression that they were cooperating, and, in fact, Muller and Bridges in particular have always freely exchanged materials and I think kept in touch with each other's work. . . . We are not interested in home runs, and above all not inclined to suspect that other scientists are going to take unfair advantage of our work." (Privately, Morgan thought Muller was probably behind Painter's outburst.) A friendly letter of explanation from Bridges finally persuaded Painter that he had been too hasty in indicting his colleagues with a moral breach.[48]

It was a tempest in a teapot, but one that offers a revealing glimpse of the strains in the drosophilists' moral economy. If the stakes were sufficiently high and institutional programs at risk, then actual behavior could depart from the ideals of cooperation and sharing. These moral ideals were designed to work in a material economy of abundance, where there were no high-stakes, zero-sum games, and where

46. T. Painter to D. F. Jones, 15 Oct. 1934, THM-CIT 1.

47. Painter to Demerec, 20 Oct. 1934, MD.

48. Morgan to Jones, 31 Oct. (quotation), 1 Nov. 1934, THM-CIT 1; and Painter to Muller, 2 Dec. 1934, HJM-C. Jones had in fact asked Bridges to referee Painter's manuscript.

Figure 5.3   The *Drosophila* group at the University of Texas, late 1920s. *Left to right:* Theophilus Painter, Clarence Oliver, Wilson S. Stone, H. J. Muller. Courtesy of Lilly Library, Indiana University, Muller Papers.

trust in the founders' goodwill could be assumed. To be sure, Painter had a notoriously suspicious and combative disposition, but he also worked in a group that was striving to achieve parity with the Caltech group.[49] Moreover, he was a newcomer to *Drosophila* work and less socialized, and he felt hard pressed by competitors (including Bridges) who had jumped on his bandwagon of cytological mapping.[50] The cytological technique was so easy to learn and the opportunities so easy to see that Painter was hard pressed to skim the cream before others did. All twelve members of the Texas group were "working literally night and day on various problems opened up by the new tool," Painter told Demerec.[51] High-stakes competition like that strained gentlemanly moral ideals.

A similar dustup occurred in 1935 between Demerec, Patterson, and a Japanese group over salivary mapping of *D. virilis*. It is a good example of how ownership rights became ambiguous as tools became

49. Maydelle Bishop to Muller, 23 Feb. 1939, HJM-C.

50. Dobzhansky to Schultz, 5 Jan. 1939, JS; Demerec to Dobhzansky, 9 Jan. 1935, MD; and W. Weaver, diary, 24 Apr. 1934, and F. B. Hanson, diary, 14 Nov. 1934, both in RF 205D 7.86.

51. Painter to Demerec, 20 Oct. 1934, MD.

objects of research. Demerec had begun mapping *virilis* in a small way as a routine preliminary to a study of its peculiar, rapidly mutating genes. Mapping was a means to a larger end, but since Demerec was the only one working on *virilis* at the time, the entire burden of mapping fell on him—he complained to Franz Schrader how it kept diverting him from what he had set out to do.[52] Then Painter's salivary technique opened up the alluring possibility of comparing genetic and cytological maps in *virilis*—the hottest line of work in town—and what had been a tool suddenly became an object. So, in partnership with Schrader (a cytologist) and Schrader's student Roscoe Hughes, Demerec plunged into systematic genetic and cytological mapping of his organism. Having single-handedly constructed *D. virilis* as a standard tool, he naturally felt that he had proprietary rights to its genetic maps.

As a result, when Patterson wrote to ask if one of his students, Wilson Stone, might use Demerec's stocks for a comparison of certain mutants in *virilis* and *melanogaster*, Demerec balked, even though Stone planned no systematic mapping. Any comparative genetics, Demerec feared, might constitute an infringement of his rights to *virilis*:

> It took over twenty years of hard breeding to obtain enough material to make this comparison now feasible. As I see it, therefore, the main part of the problem is not to make a comparison when the material is already available but to get the material for the comparison. The situation with virilis is similar to that with simulans. In case of simulans the genetic analysis of that species was one man's job (Sturtevant's) and therefore, though the mutations found by him are available to anyone wishing to use them in work on special problems, he is the person to make comparisons between chromosomes of melanogaster and simulans.[53]

He felt he had a similar claim on *D. virilis*. If the Texans thought they could extend or improve what Demerec and Schrader were doing, they were free to do so, but then they could also do the hard work of getting additional mutants.

Schrader, too, worried that the aggressive Texans would take over their project if given half a chance. Indeed, it seemed to him that they were taking over just about everything else:

> I move that you [Demerec] and the Pasadena people—in fact every Drosophila geneticist and cytologist—give up working on Drosophila and

52. Demerec to Schrader, 6 Mar. 1929, MD.

53. Demerec to Patterson, 22 Jan. 1935, MD; see also Patterson to Demerec, 18 Jan., 8 Feb. 1935, MD.

Figure 5.4   Franz and Sally Schrader, cytologists and close friends of the Columbia drosophilists, late 1920s. Courtesy of American Philosophical Society Library, Stern Papers, photographs file Schrader.

put everything into the hands of the Texas geneticists. I'm not sure but that it may be advisable to let the University of Texas take over all plant genetics also.

P.S. I'll be glad to send you a list of cytological problems that have no genetic implications. I stand ready to help you, especially in regards to cytological technique, and feel sure that you will soon feel at home in my side of the field. I hope you won't feel that you will have to quit biology altogether.[54]

Demerec did finally send Patterson the stocks, but reluctantly.

Demerec and Schrader's fears were in fact realized, not by the Texans but by a new group of drosophilists at the University of Kyoto: H. Kikkawa, T. Komai, and S. Fujii, who were building a large program around *D. virilis* and some local species. Demerec was aware that the Japanese were mapping *virilis* (they had exchanged stocks) but had understood that they were doing so only as a preliminary to a special study of certain translocations and inversions. So he and Schrader were taken aback when, in 1936, Kikkawa informed them that they were

54. Schrader to Demerec, 24 Jan. 1935, MD.

about to publish complete cytological maps of *virilis*. Schrader felt double-crossed, but Demerec reminded him that they had agreed that both groups would proceed with the cytological mapping, since the method was of such wide interest, many groups would be doing it. Demerec faulted the Japanese only for not keeping them better informed of their progress.[55] A similar but much more bitterly contested dispute erupted in 1939 between Caltech and Cold Spring Harbor over the rights to complete and publish Bridges's cytological maps of *D. melanogaster*, which he left unfinished at his death.[56]

It is no accident that these displays of competitive behavior were all inspired by cytological mapping. Genetic mapping had by the mid 1930s become largely a tool. However, Painter's technique of cytological mapping transformed genetic maps from tools into objects of systematic comparative mapping, cytological and genetic, in *melanogaster* and other species (figure 5.5). This kind of comparative work was easy to do and very highly valued, since it was the first time that "distances" defined by linkage could be given a physical meaning. Maps were such fundamental tools, used by every drosophilist in every project, that whoever succeeded in producing standard cytological maps was virtually certain to enjoy the same visibility and prestige that Bridges and Sturtevant had enjoyed in the 1910s. Producing standard maps would make a group's fame, but the accessibility of the salivary method and the number of competing groups made the opportunity fleeting and hotly contested.

There had been no question about Bridges's and Sturtevant's proprietary rights to the mapping of *melanogaster*, since they had had no competitors. In the late 1930s, however, those who invested in systematic mapping and constructing standard flies could no longer count on reaping the rewards of their labors. Customs of cooperation and avoidance that were designed for a world with one leading center were strained in a more competitive world. Relations between Texas, Caltech, and Cold Spring Harbor remained cordial, but wary. Competition became more unpredictable, and competitive feelings became

55. Schrader to Demerec, 26 Feb. 1936; and Demerec to Schrader, 29 Feb. 1936; MD.

56. Demerec to Morgan, 9 Feb., 15 Apr., 14 Nov. 1939; Morgan to Demerec, 15 Feb., 23 Oct., 17 Nov. 1939; Morgan to V. Bush, 24 Apr. 1939; Demerec to W. M. Gilbert, 23 Jan., 10, 20 Nov. 1939; and Demerec, "Memorandum on unfinished problems of C. B. Bridges," 23 Jan. 1939; all in CIW file Bridges. It was the old rivalry between the two Carnegie fly groups that made the dispute so bitter.

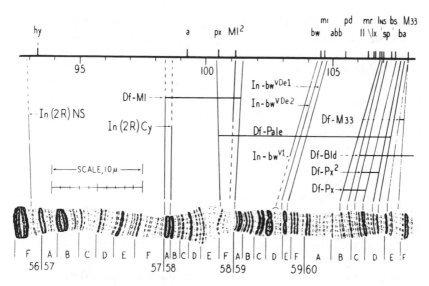

Figure 5.5 Linkage and cytological maps of a salivary chromosome, showing homologies. From Calvin Bridges, "Correspondences between linkage maps and salivary chromosome structure, as illustrated in the tip of chromosome 2R of *Drosophila melanogaster*," *Cytologia* (1937) Fujii Jubilee: 745–55, fig. 1, p. 750.

harder to regulate by collegial customs of reciprocity, disclosure, and communal ownership of tools.

Formalizing the Exchange Network

As the community of drosophilists became larger and more diverse, the exchange system network lost some of the personal, familial quality that it had in its first decades, when Columbia was the only center of supply. In 1934 a second official center of stock distribution was established at Cold Spring Harbor, and the first issue appeared of the *Drosophila Information Service*, a mimeographed technical newsletter jointly produced by Demerec and Bridges. These new services were the brainchildren of Demerec and were part of his effort to make Cold Spring Harbor an international center to rival Caltech.

There were many reasons why new forms of community organization emerged in the mid 1930s, but one important reason was the hitch in the supply of stocks that occurred when the fly group migrated to California in 1928. Shifting the center of supply west by 2,500 miles caused real problems of supply. Commercial air mail was still exotic and expensive, and long transcontinental train trips exposed drosophilas to the hazards of torrid and frigid weather. Shipping to Europe from California was even riskier. It became obvious, to Demerec at least, that a second stock center was needed in the eastern United States. [57] More precisely, Morgan's removal to the Pacific slope created an opportunity for Demerec to do what he and Metz had dreamed of doing since 1919—to create a second fly group at Cold Spring Harbor.

In fact, Morgan took the occasion of his move west to shed some of the burden of supplying stocks to an ever-expanding network of users. He arranged to transfer part of the supply operation to G. M. Gray, the keeper of animal stocks at Woods Hole. Exactly how much of the operation Morgan intended to leave behind is not clear. He told L. C. Dunn that "the whole business" had been shifted to the Marine Biological Laboratory, but apparently no more than a small number of stocks were ever kept there, just those most frequently used. Most likely Morgan meant Gray to provide only the most commonly requested stocks.[58] In any case, the arrangement was disastrous. Gray was already overworked, and since Woods Hole was not a year-round center of *Drosophila* research, Gray had no incentive to keep up stocks. He confessed that there were "times of neglect," but the situation was worse than that: when Dunn inquired about it in 1930 he found just one lone culture left alive![59]

In hindsight Morgan would have been wiser to trust the Carnegie stocks to Cold Spring Harbor, where it would have been linked organically to Demerec's active research program. The reason he did not, obviously, was the ancient rivalry between the two Carnegie labs. The then CIW president, John C. Merriam, was already worrying that the

57. Demerec to V. Bush, 27 Feb. 1939, CIW file Drosophila Stock Centers.

58. L. C. Dunn to G. M. Gray, 11 Feb. 1930, LCD; and Dunn to F. Lutz, 20 Oct. 1928, Dunn to H. Plough, 14 Oct. 1930, both in LCD file NRC Committee on Experimental Animals and Plants (hereafter, file NRC). Demerec alleged in 1939 that it was hard to get stocks from the Morgan group, but he may have been speaking in anger: see Weaver to Bush, 10 Apr. 1939, and Bush to Weaver, 11 Apr. 1939, both in CIW file Drosophila Stock Centers; and Schultz to Demerec, 13 Mar. 1937, MD.

59. G. M. Gray to Dunn, 13 Feb. 1930, LCD file NRC. A quasi-commercial supply house operated successfully out of the marine biology station at Pacific Grove, but it had connections with an active research group: A. E. Galigher to "professor," 20 Apr. 1923, CZ 7. Commercial supply houses that lacked a base in an active research program were notorious for frequent errors in identification and quality control: C. S. Gager to Dunn, 24 May 1930, LCD file NRC.

Carnegie team would lose its identity at Caltech, and the Carnegie stocks constituted the single most compelling reason, in Morgan's view, why the CIW should continue to support his group.[60] Morgan would hardly give Merriam a reason to shift support to his rival, Demerec.

When the Woods Hole supply scheme failed, the burden of supplying stocks fell on the smaller centers of *Drosophila* work in the East. Columbia continued to get requests for stocks, even though Donald Lancefield was the only drosophilist left there. As Dunn reported, his laboratory "in common with a great many others has been in receipt of continual requests for stocks." Harold Plough at Amherst also felt the burden: "Since Professor Morgan stopped furnishing these cultures," he complained, "I have sent out a fair number, and requests seem to be increasing. Unless some satisfactory supply for these cultures is developed, I am afraid that I too will have to discontinue the practice." He was willing to hire an assistant and charge fees to cover costs but thought the traditional free exchange was better.[61]

Nothing more was done until January 1934, when some leading drosophilists got together to consider the problem of supply. They first agreed that the five largest eastern labs would between them maintain and make available 450 stocks. This plan was shelved, however, when it transpired that the Rockefeller Foundation and the CIW were interested in creating a second stock center for the eastern United States at Cold Spring Harbor. CIW president Merriam agreed to underwrite the center, and Bridges, who was delighted to have someone with whom to share the burden of supply, offered to send Demerec whatever stocks he wanted.[62] Demerec was of course pleased to have his laboratory at the center of activity, and before long Cold Spring Harbor had a stock collection and a network of supply that rivaled Caltech's. In its first eight months 541 cultures were sent or received, and 340 in 1935.[63]

Morgan was not pleased with this development, as he told Demerec in no uncertain terms:

60. J. C. Merriam to Morgan, 3 May, 23 Oct. 1928, Morgan to Merriam, 11 May, 5 Nov., 14 Dec. 1928, all in CIW; and Merriam memos, 27 Sept. 1934, 4 Jan. 1935, CIW file Merriam Memos Genetics.

61. Dunn to E. B. Babcock, 15 Oct. 1930; and H. Plough to Dunn, 30 Sept. 1930; LCD file NRC.

62. Demerec to Dunn, 11 Jan. 1934; Demerec to Bridges, 10 Apr. 1934; Bridges to Demerec, 14 Apr. 1934; all in MD.

63. Demerec. "Memo on activities involving international cooperation," n.d. [1936], MD file Blakeslee.

Figure 5.6 Milislav
Demerec with a salivary
chromosome map at sum-
mer symposium, Cold
Spring Harbor, 1941. Courtesy of American Philosophical Society Library, Dunn
Papers, small photo album.

> Morgan said to me that I had no business to accept any money for Dro-
> sophila work; if any support for this work is forthcoming it should go to
> his laboratory and not to any other place, and he particularly objects to
> having the work on Drosophilia developed at Cold Spring Harbor since
> he feels that may jeopardize the grant he is receiving from the Carnegie
> Institution. Morgan was very frank. He probably expected that I would
> stop working with Drosophila.[64]

Morgan probably did not mean for Demerec to drop *Drosophila*, but he
certainly did not welcome a second Carnegie stock center. But there
was nothing he could do. The fly people had simply become too numerous and diverse to rely on a single source of stocks and leadership.

A growing international traffic in *Drosophila*—a consequence of
the many new *Drosophila* labs in Europe, the Soviet Union, and Japan—
created special problems for an informal exchange system. To the fly
people drosophilas were harmless benefactors of humankind, but to

64. Demerec to W. M. Gilbert, 10 Nov. 1939; see also Morgan to Gilbert,
31 Oct. 1939; both in CIW file Bridges.

fruit growers and agricultural bureaucrats they were pests, and under the laws regulating trade in agricultural products it was illegal to transport live drosophilas across international boundaries. (It was also illegal to transport them within the United States, at least without a special permit; but since postal inspectors almost never opened domestic mail, there was no risk of interdiction.) Many drosophilists had stories to tell of customs agents destroying valuable collections of mutants, deaf to every plea. The Russian drosophilist Israel Agol lost his stocks on entering the United States for a visit in 1930, and warned Stern that if he were coming to the States he had better make the entry of his flies "as unofficial as possible." The Rockefeller Foundation tried to intercede but to no avail. Muller's solution was to send duplicate sets of stocks through the international mails; since customs inspectors opened only one package in three there was a good chance of one getting through.[65] Bridges, in contrast, just used the old tourist trick when he came back from Europe in 1932: "I smuggled all the stocks in without difficulty. They were concealed in my dirty clothes and weren't seen at all."[66]

Informal evasions could not be employed, however, with an operation the size of Demerec's, so after trying in vain to persuade the bureaucrats that *Drosophila* was not a pest, Demerec applied for and ultimately got a permit from the Department of Agriculture to import and export live flies.[67] It struck drosophilists as funny that the government should want to regulate the travels of their inveterate hitchhiker, and jokes began to circulate about "Drosophila immigration visas." The human migration from Europe in the late 1930s, which included drosophilists like Richard Goldschmidt, gave the joke a sharper edge. Alvin Johnson, who led efforts to get Jewish scientists out of Germany, quipped: "I thought we had vinegar flies enough in this country, and as a hundred-percent American I have my reservations about immigrant Drosophila." (Demerec missed the joke.)[68]

Demerec clearly felt he was continuing the honored tradition of the *Drosophila* exchange system, and in many ways he was. But some of the personal quality of the tradition was lost when Bridges ceased to be

65. Muller to Stern, 25 Sept., 16 Oct. 1930, 4 Feb. 1931, CS; and Demerec to Bush, 8 Apr. 1939, CIW file Drosophila Stock Centers.

66. Bridges to Stern, 1 Apr. 1932, CS.

67. Demerec to L. A. Strong, 4 Apr. 1934; Strong to Demerec, 2 May 1934; Demerec to Gilbert, 1 June 1934; Demerec, "Report on shipments," 1 Feb. 1940; all in CIW file Drosophila Program.

68. A. Johnson to Dunn, 16 Feb. 1941, LCD; Demerec to Johnson, 17 Feb. 1941, LCD; and Muller to Demerec, 24 Oct. 1939, 24 June 1940, MD.

the sole source of supply. It became less an exchange of professional favors than a not-for-profit service.

The *Drosophila Information Service*

If stock centers were the material technology of the *Drosophila* exchange system, the *Drosophila Information Service (DIS)* was its literary or communication technology. The *DIS* was Demerec's idea and was modeled on the mimeographed newsletter that Demerec's teacher at Cornell, Rollins Emerson, distributed to maize geneticists to keep them up to date on the progress of the genetic mapping. However, it was coproduced with Bridges and was very much a cooperative venture.[69] The *DIS* was, in effect, an informal trade journal; it used publicity and information to give the fly people access to the news of the profession and to new technical developments while at the same time gently enforcing standard practices.

There was, in fact, a growing need for more rationalized communication as *Drosophila* workers proliferated in the early 1930s. The system was becoming too big for everyone to know what everyone else was doing through personal acquaintance. For twenty years the predominance of the Morgan fly group ensured a degree of uniformity and standards: Bridges's practices were de facto the standard practices, because he had made the maps and invented the experimental procedures that everyone used. Everyone knew what the standard practices were because they were taught or supplied by Bridges. That ceased to be the case as the system expanded. Many second- and third-generation drosophilists had not learned their tradecraft from the original source. New techniques like X-ray mutation and salivary chromosome cytology compounded the problem of communication by making it extremely easy for anyone, anywhere, to produce rich experimental material.

Inconsistent naming of mutants became a real problem when Muller's X-ray method began to produce its deluge of deletions, inversions, and other chromosomal aberrations after 1928.[70] Decentralization of production also made it harder for drosophilists to agree on standard names. "Due to the lack of contact between labs," Demerec recalled, "identical changes were frequently given different names, or different

69. Bridges to Demerec 1 Oct. 1933, MD; Demerec to Bridges, 9 Oct. 1933, MD; and Bridges and Demerec to *Drosophila* geneticists, 10 Nov. 1933, HJM-C. Demerec was sometimes exasperated by Bridges's procrastination, however: Demerec to Bridges, 13 Feb. 1934, MD.

70. Demerec to Schultz, 6 Jan. 1934; Patterson to Demerec, 22 Sept. 1933; and Demerec to Patterson, 30 Aug. 1933; all in MD.

Figure 5.7 Calvin Bridges in the Caltech *Drosophila* stockroom, circa 1935, with the first issue of *Drosophila Information Service,* containing his catalogue of "standard" *melanogaster* mutants and map locations. Courtesy of California Institute of Technology Archives, Photo File 10.24.

changes were given the same name; systems of nomenclature were being modified to suit the needs of the individual laboratories, without regard for the need of the Drosophila group as a whole.[71] One of the first things that Demerec and Bridges agreed to do was to draw up a standard list of names and symbols for mutants. Bridges got that job, of course—"a GREAT BIG JOB," he began to realize when halfway into it.[72]

Problems of access to standard methods were behind a second feature of the *DIS,* its technical notes on everything from the design of incubators and etherizers to improved methods of collecting *Drosophila* eggs and remedies for epidemics of mites.[73] This service made up for deficiencies in the more informal, traditional methods of communicating craft knowledge. Access to the nitty-gritty of experimental work had always been largely a matter of tacit knowledge, acquired through apprenticeship in centers of production or by personal correspondence. It was not usual at the time to publish the details of experimental methods; Bridges did from time to time, but usually long after the methods

71. Demerec to V. Bush, 8 Apr. 1939, CIW file Drosophila Stock Centers. Cf. *DIS* 11 (1939): 5–6.

72. Bridges to Demerec, 8 Feb. 1935, MD. See also Bridges to Demerec, 1–4, 22 Apr. 1935; and Demerec to Bridges 14 Feb. 1935; all in MD.

73. *DIS* 6 (1936) was devoted to research methods.

had come into common use through informal learning. The technical notes of the *DIS* were intended to make all the little improvements in the craft of *Drosophilia* work immediately accessible to everyone, well connected or not, and to make newcomers to the fly less dependent on personal connections.[74] Up-to-date directories of *Drosophila* laboratories and their staffs also made it easier for newcomers to find out who was who.

A third and most important service of the *DIS* was to publish lists of the stocks that were available from every working *Drosophila* lab. Updated every year or so, these stock lists were shopping lists for all drosophilists: a reliable and accessible directory to the new, multicentered supply system—its information software, so to speak. Caltech, Cold Spring Harbor and Texas had by far the most complete collections, (825, 442, and 417 stocks, respectively, in 1939), though Texas could not ship stocks for much of the year owning to local extremes of heat and cold. Smaller laboratories served local or regional markets or supplied local varieties or special stocks, which the major centers would not bother to maintain.[75] (The number of local varieties and species in use increased dramatically in the late 1930s.)

Virtually every feature of the *DIS* was a formalized version of a service that Bridges had provided informally. It had long been Bridges's custom, for example, to circulate a mimeographed list of stocks that he could supply, as well as mimeographed forms that drosophilists could use in requesting information on mutants or linkage data. The *DIS* stocklists were simply an expansion and rationalization of that personal service.[76] Likewise, Bridges was preparing to publish under *DIS* auspices the comprehensive *Drosophila* bibliography that he had compiled for his own use, when he learned that Muller would do it instead (he was delighted to be relieved of that chore).[77] The technical notes published in the *DIS* were a more systematic version of Bridges's famous scribbled letters and postcards, reproduced and distributed for all the fly people to use, not just personal friends and colleagues. The *DIS* directories broadcast gossip that had previously circulated by personal letters.

As a literary genre, the *Drosophila Information Service* combined the

74. Demerec to V. Bush, 8 Apr. 1939, CIW file Drosophila Stock Centers.

75. Ibid.; and stocklists in *DIS* 10 (1939).

76. Bridges to Demerec, 1 Oct. 1933, MD.

77. Muller to Demerec, 7 Apr. 1935; Demerec to Muller, 19 Apr. 1935; and Bridges to Demerec, 22 Apr. 1935; all in MD.

qualities of a personal letter and an official journal. As a result, it had an ambiguous status as a public document and raised awkward problems of proprietary rights. Demerec and Bridges wanted to encourage drosophilists to exchange information as freely in the *DIS* as they would in private letters. Yet broadcasting private knowledge to everyone without some protection of property rights might invite poaching and inhibit free exchange. Demerec and Bridges attempted to solve this dilemma by displaying prominently on the cover of the *DIS* the statement, "This is not a publication." It was not a surrealist manifesto (like Magritte's "Ceci n'est pas une pipe") but a marker of the uncertain boundary between private and public knowledge. It meant that single items of information could be appropriated freely but that larger chunks, such as linkage maps, could not be republished without giving personal credit.[78] Demerec's creative use of ambiguity echoed a line with which Sturtevant had opened a letter to him in 1930: "This is not a letter but only an announcement of a new mutable gene."[79] It was one of Sturtevant's little jokes, but it reminds us that personal letters also served a quasi-public purpose as a medium for communicating news and know-how. In the fly group, certainly, letters from colleagues were read by everyone and publicly discussed; thus, drosophilists had some experience in dealing with ambiguities in the moral economy of exchange. Perhaps that is why the *DIS* worked so well as a formal public medium, which preserved some of the informality of private correspondence.

The *DIS* was about as far as the fly people were willing to go in centralizing authority, however. They flatly refused to set up an international committee to ordain rules for standard nomenclature; they preferred to give the task to Bridges, whom they knew and trusted.[80] When the editor of *Genetics*, Donald Jones, tried to impose standard terms and abbreviations, he was ridiculed by Sturtevant, who gleefully conjured up a paper "by Sturt and Doby on the chroms of Dros mel, in which shall be references to the work of Morg, Mull, Paint, and Patters." Better, he thought, to let authors use their own judgment "and trust natural selection to do the rest."[81] Similarly, Theophilus Painter quashed Demerec's suggestion that the groups at Texas, Caltech, and Cold Spring Harbor collaborate on salivary mapping:

78. *DIS* 2 (1934): 3–4.
79. Sturtevant to Demerec, 15 Mar. 1930, MD.
80. Demerec to Bridges, 2 Jan. 1934, MD.
81. Sturtevant to D. F. Jones, 3 Aug. 1934, AHS 2.18.

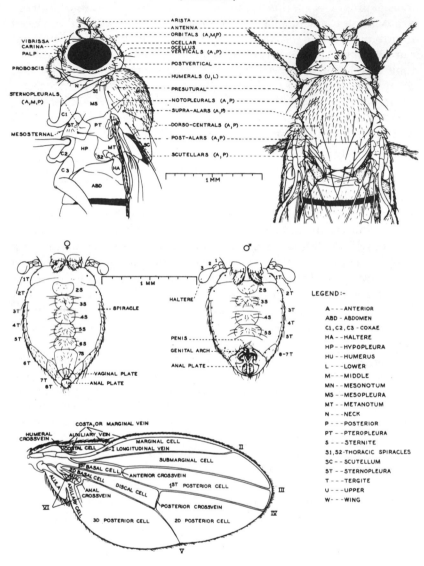

Figure 5.8 Calvin Bridges's standardized *Drosophila* worksheet, devised as an aid to uniform identification and naming of mutants. *Drosophila Information Service* 6 (April 1936): 76.

Here in America, open competition in research as in business is the most progressive policy and . . . any attempt at the regimentation of research is doomed to failure. As we see it, with our set-up of competing institutions . . . it would be useless and harmful to the progress of our science for any one institution, or group of individuals, to attempt to regulate the scientific research in any one field. The major problems are perfectly obvious to any one who knows the field and for the next year or two I think it highly desirable that we should all range out, as it were, and work on those phases which interest us most."[82]

Charles Metz agreed: sharing tools was a good thing, but collective control of individuals' choices was going too far.[83]

In a multicentered system of more or less equal producers, customs of personal trust were bound to erode. Older drosophilists especially were aware of this subtle change in the emotional feel of their collective life. Bridges and Demerec noted with regret that "the intimate contact which existed between the Drosophila workers of the past has been lost."[84] They expressed this sense of loss in technical rather than emotional terms. They worried, for example, that the loss of personal connections would result in duplicated effort and a loss of unpublished but still useful information, which before had been kept in public circulation in personal correspondence. But clearly more was at stake than efficiency of production: a way of life, a distinctive culture of practice was also changing, a distinctive way of interacting and a distinctive moral economy. A new generation of drosophilists was coming into its own, whose members were more diverse and competitive and took for granted their possession of the means of production and their free access to choice problems. The original fly group could no longer set the standards and the agenda for all the fly people. The system was just too big and too dispersed. Demerec and Bridges meant the *DIS* to preserve the emotional and moral qualities of the drosophilists' community of practice, and to an extent they succeeded, but it was no longer quite the same community.

82. Painter to Demerec, 20 Oct. 1934, MD; see also Demerec to Painter, 24 Oct. 1934, MD.

83. Metz to Demerec, 25 Oct. 1934, MD.

84. Bridges and Demerec to *Drosophila* geneticists, 10 Nov. 1933, HJM-C; cf. *DIS* 11 (1939): 5–6.

Conclusion

The drosophilists' exchange and communication network was the third element in their system of production. Standard fly, fly group, and exchange system were their material, social, and literary or communication technologies. This network of connections and customs of free exchange, no less than customs of cooperative benchwork, enabled drosophilists to exploit the enormous productive potential of their breeder reactor. It embodied the drosophilists' defining social and moral principles, and created standard practices and a distinctive way of experimental life.

The exchange system and its moral rules came into existence at a time when many drosophilists were small-scale, widely dispersed producers with minimal resources. It began as a bootstrap system of self-help designed for conditions of general scarcity, and its customs and moral values became traditions embedded in the communal work habits of the fly people. While the growth in the number of practitioners made the system more impersonal and competitive in the 1930s, these changes did not fundamentally alter the customs and values preserved in the system of stock exchange. It was these rules of exchange, as well as the rituals of group work, that gave the fly people their distinctive identity. These were defining traits in their view: the fly people were a people who took as standard the tools and practices of their founding group. They were a people who discouraged monopolies, shared tools and trade secrets, and did not poach and ambush. The communal summer researches at Woods Hole and other marine experiment stations (also institutional relics) sustained similar values in other communities of experimental biologists. However, the sense of communal identity seems to have been especially strong among the fly people, no doubt because of the unusually strong and durable authority of the founding group and its ability to sustain a high level of productive work in an increasingly competitive world. Standard flies were not just the means of experimental production but also the bearers of a distinctive moral economy and a distinctive way of experimental life.

From a natural history perspective, the exchange system was the principal means of dispersal of domesticated drosophilas, the key to the coevolutionary relation between humans and flies. Standard *Drosophila*, like any new kind of creature, was created locally and could become cosmopolitan only via some effective means of dispersal. In the wild, *D. melanogaster* had spread worldwide by attaching itself to humankind, following the spread of agriculture, viticulture, and the fruit-and-vegetable trade. Similarly, standard *melanogaster* stocks spread around the scientific world via the fly people's exchange network—a most convenient and effective vehicle for that inveterate hitchhiker. Stock exchange was a selective system of dispersal, and one quite different from dispersal in nature. The mutants that traveled and established themselves most widely were not the hardiest but the ones most useful for doing experiments and producing papers. Standard flies traveled not on the summer breezes or as stowaways in fruit boats, but in culture tubes via the public mails. Through sleet and rain, across state and (with luck) national boundaries—always well-provisioned, always more or less indoors. This mobility is what made the standard fly standard.

For drosophilists, too, the exchange system was a vehicle of dispersal and colonization and transformed them from a local species of experimental biologists to a cosmopolitan one. Availability of standard flies made every laboratory, however small and out of the way, potentially a place where drosophilists might flourish and reproduce their kind. The exchange network extended the ecological range of the fly people's practices and distinctive way of life by enabling small, dispersed producers to participate in some measure in the research front. Doubtless research on *Drosophila* genetics would have been carried out without such a system of dispersal, but it would arguably have been less expansive intellectually, and it would certainly have spread more slowly.

Creating major centers of *Drosophila* work within this dispersed network required more than standard stocks, however. The limiting factor for institution building was not access to standard tools and craft knowledge but access to institutional resources. If the local ecology was favorable, the exchange network enabled individuals to flourish on minimal means. But institution building depended on local allies and resources, and the scarcity of such resources is doubtless one reason why the original fly group had so few rivals for so long.

What caused new fly groups to take shape in the late 1920s and 1930s was nothing internal to *Drosophila* genetics but large-scale changes in institutional systems: to wit, the wider acceptance of research as a constitutive part of academic careers, and new sources of research funds from public and private agencies, most notably the Rockefeller and Rothschild Foundations, the Kaiser Wilhelm Gesellschaft, the Soviet Academy of Sciences, and the ministries of education in France and Japan. No fewer than four of the ten or so major new *Drosophila* centers in the late 1930s owed their existence to Rockefeller funding: the Rothschild Institute in Paris, and the Universities of Edinburgh, Texas, and Rochester. In the 1930s more academic zoology departments were eager to support research projects, and the existence of extramural patrons and the rise of a new generation of trained, bright drosophilists were strong incentives to entrepreneurs like Demerec, Stern, Patterson, Boris Ephrussi, N. W. Timofeeff-Ressovsky, and F. A. C. Crew (the founder of the Edinburgh school of drosophilists) to create fly groups on the model of Morgan's original. Access to standard tools and methods via the exchange network was an essential part in creating infrastructure, but it was a necessary and not a sufficient cause in spreading the fly people's distinctive way of life around the world.

PART II

Expanding the System

SIX

Improvisations

THE INCORPORATION OF *DROSOPHILA* into a system for producing genetic maps created a mode of practice that was far more dynamic and productive than older styles of experimental heredity, but also narrower. The big problems of experimental biology—variation, evolution, development—were set aside when Morgan and his group made the topography and behavior of chromosomes their main goal. *Drosophila* genetics expanded and diversified in the 1910s and 1920s around a cluster of fundamental concepts and problems that derived from the mapping business: the physical nature of the gene and gene action, the process of crossing-over, the nature of mutation, and genic interactions. These key ideas dominated theoretical debate and structured empirical experimental work.

This definition of the mainstream is evident in the projections of future work that Morgan sent to his Carnegie patrons. The nature and causes of mutation, the cytology and physiology of crossing-over, sex determination, the extension of linkage groups to other species— these were the big problems for drosophilists in Morgan's view, in addition to filling in the maps.[1] The new inventions that drosophilists valued most highly in the 1920s were those that could be used to probe the machinery of crossing-over—Lillian Morgan's attached-X mutant, in which two X chromosomes were joined end to end; Bridges's triploid mutant and his series of intersexes and supersexes; H. J. Muller's suppressors of crossing-over; Muller and Bridges's deletions, duplications, and translocations; and so on. Such "trick chromosomes" were at first challenging genetic puzzles for *Drosophila* virtuosos. However, they soon became specialized tools for analyzing the mechanics of genetic transmission. Leading second-generation drosophilists—Milislav Demerec, Curt Stern, N. W. Timofeeff-Ressovsky, Otto Mohr, and others—made their reputations in the mainstream with work on mutation, genic interactions, dosage effects in haploids and polyploids, and so-

1. T. H. Morgan, "Program of work," to R. S. Woodward 2 Oct. 1917, and Morgan to J. C. Merriam, 14 Dec. 1928, CIW.

matic crossing-over.[2] In short, chromosomal mapping generated a set of rich and leading theoretical problems for drosophilists to explore, and a diverse array of versatile, specialized experimental tools and methods. These problems and methods defined for a generation or two what *Drosophila* genetics was, and how careers could be made doing it.

The productivity and versatility of chromosomal mechanics tended to discourage other lines of work. Maternal or cytoplasmic inheritance became somewhat of a heresy for core drosophilists, for example, because it seemed to threaten the central role of chromosomes and genes. Materially this problem became more difficult to investigate because the drosophilists' experimental tools had been designed and constructed to study chromosomes. There was no pigeonhole for cyto-genes in the drosophilists' accountings system, no place for them on genetic maps. Maternal inheritance was simply not provided for in the blueprint of standard *Drosophila*. Nor did it have a place in the social design of drosophilists' community of practice. Nonchromosomal inheritance was more likely to be perceived as a threat to be contained than as an enrichment to be assimilated.[3]

Similarly, it became more difficult to use *Drosophila* for probing the hidden machinery of individual development or the buried history of species. These aims were not rejected because they were threatening or heretical; they were just less productive than mainstream lines and thus could not compete for drosophilists' attentions. Conceptions of evolutionary and developmental genetics were less clear-cut than conceptions of chromosomal mechanics, and experiments designed to illuminate development and evolution produced fewer theoretically significant and usable results than did mainstream investigations of crossing-over. Such work won less attention and was less likely to result in upwardly mobile careers. Few drosophilists would have said that the genetics of development and evolution was not important: quite the contrary. But when they tried to use *Drosophila* to produce such knowl-

2. Elof A. Carlson, *The Gene: A Critical History* (Philadelphia: Saunders, 1966), chaps. 12–15.

3. Jan Sapp, "The struggle for authority in the field of heredity, 1900–1932: New perspectives on the rise of genetics," *J. Hist. Biol.* 16 (1983): 311–42; Sapp, "Inside the cell: Genetic methodology and the case of the cytoplasm," in J. A. Schuster and R. R. Yeo, eds., *The Politics and Rhetoric of Scientific Method* (Dordrecht: Reidel, 1986), pp. 167–202; Sapp, *Beyond the Gene* (New York: Oxford University Press, 1987); and Jonathan Harwood, "The reception of Morgan's chromosomal theory in Germany: Inter-war debate over cytoplasmic inheritance," *Medizinhistorisches Journal* 19 (1984): 3–32.

edge, they usually were disappointed in the results and withdrew to more productive and bankable work on chromosomal mechanics.

The purpose of this and following chapters is to understand how capacities and limitations were built into the *Drosophila* system and affected the work that drosophilists did. I do not deal in this book with the varieties of mainline chromosomal mechanics, which are already more or less familiar, but focus instead on drosophilists' efforts along the margins of their field, along the boundaries with development and evolution. It is along these frontiers that the constructed nature of *Drosophila* is most visible, because it was there that drosophilists were attempting to use their instrument for purposes for which it had not been designed. When drosophilists pushed their machinery of production beyond its limits, historians can catch glimpses of how material and social mechanisms define experimental practices, and how deviant practices were reabsorbed into the mainstream.

Drosophila, like most instruments, is a multipurpose technology, and its versatility encouraged improvisation. Yet the improvisations tended to falter and revert to mainstream chromosomal mechanics. Trick systems such as intersexes, triploids, and attached-X mutants were designed for studying the mechanism of crossing-over, but they could also be diverted to the study of development. Indeed, in the absence of tools specifically designed for the genetics of development, drosophilists had every incentive to improvise with tools already to hand and of proven value. Precisely how such trick systems would end up being used was indeterminate. Work done with them, however, for whatever initial purpose, tended to become chromosomal mechanics. That was what the *Drosophila* system had been designed to do and what drosophilists knew how to do best. Their multipurpose technology offered drosophilists a fallback position when risky improvisations seemed not to be working out. The indeterminate nature of the *Drosophila* system and a strategy of improvisation altered the calculus of risk, making it easier for experimenters to revert than to persist in a risky venture.

The capacities built into experimental tools like *Drosophila* do not determine their uses in any simple deterministic way—experimental behavior is much too flexible to be so constrained. Nor did a *Drosophila* establishment enforce a party line or punish deviance—experimenters were far too individualistic and independent to be so disciplined. There are many reasons why systems of production like *Drosophila* develop a momentum or inertia: the need to get a full return on an initial investment and avoid the disruptions of retooling, the visceral pleasure of exercising familiar skills, or the desire to preserve privileges and power relations that are built into successful ways of making things.[4] We are dealing here with neither a material nor a sociological determinism, but with a material and social system of production, the design and accumulated experience of which inclined experimenters' choices in particular directions. There was freedom of choice, but the odds were stacked. So while programmatic ambitions to reconnect genetics with other branches of biology impelled drosophilists to improvise new variants of practice on the borders of development and evolution, their previous investments in skills and know-how, together with the practical need to produce recognizably significant work, drew innovators back to proven mainstream lines. It was the practical realities of the production process that winnowed variability in experimental behavior and sustained dominant practices.

I will focus in this chapter mainly on the fly group because they, I think, did the most sustained and ingenious work in trying to reconnect *Drosophila* genetics with development and evolution. This may seem an odd claim to readers familiar with revisionist histories of the Morgan school. The received wisdom nowadays is that the fly group eschewed developmental and evolutionary problems and used their social authority to promote a narrowly reductionist conception of genetics. They became, in this view, a "hegemonic" establishment, from which genetics had to be liberated by people with a broader biological outlook.[5] It is true that the Morgan crowd did sometimes sound like prophets of a new religion. Even well-wishers like Franz Schrader chided them for being too dogmatic and quick to dismiss more holistic biological approaches like that of Richard Goldschmidt.[6] On the other hand, Morgan wrote and published continually and copiously about embryology and evolution: five books in addition to many lectures and

4. Thomas P. Hughes, "Technological momentum in history: Hydrogenation in Germany, 1898–1933," *Past and Present* 44 (1969): 106–32.

5. Sapp, "Struggle for authority" is a classic statement of this view; see also Greg Mitman and Anne Fausto-Sterling, "Whatever happened to *Planaria*? C. M. Child and the physiology of inheritance," in Adele E. Clarke and Joan H. Fujimura, *The Right Tools for the Job: at Work in Twentieth-Century Life Sciences* (Princeton: Princeton University Press, 1992), pp. 172–96. For general background see Garland E. Allen, "T. H. Morgan and the split between embryology and genetics, 1910–35," in T. J. Horder, J. A. Witkowski, and C. C. Wylie, eds., *A History of Embryology* (Cambridge: Cambridge University Press, 1986), pp. 113–46; and Klaus Sander, "The role of genes in ontogenesis—Evolving concepts from 1883 to 1983 as perceived by an insect embryologist," ibid., pp. 363–95.

6. F. Schrader to C. Stern, 30 June 1928, 14 Mar. 1931, CS.

review articles between 1915 and 1935.[7] The subjects, it seems, were always on his mind. It is also true that the fly group did not publish a great deal on developmental and evolutionary genetics before the mid 1930s. However, the published record is not a reliable indictor of what actually was done. In fact, Sturtevant, Jack Schultz, and Charles Metz made substantial investments in developmental and evolutionary genetics, and they kept coming back to these problems despite repeated disappointments. It is no accident that George Beadle, Boris Ephrussi, and Theodosius Dobzhansky invented new systems of developmental and evolutionary genetics while in the fly group: their achievements were the fruits of years of efforts by Sturtevant, Schultz, and others to improvise ways of putting genes back in the whole organism.

It was not lack of interest or hegemonic ambitions that kept drosophilists from doing more in the 1910s and 1920s in developmental and evolutionary genetics. It seems, rather, that they were constrained by the difficulty of inventing experimental systems that could compete with mainstream production. The revisionist cultural-political historiography favored in the 1980s needs to be revised to take into account the material culture of experimental practice. We need to be less quick to impute to historical actors the contentious cultural politics of our own puritanical and sectarian age.

Genetics AND . . .

For Morgan the separation of chromosomal mechanics from development and evolution was a matter not of principle but of practical expediency. He sympathized with critics who said that a genetics of transmission without reference to development was "no more intellectual than . . . a game of cards." He did believe that the two fields should be temporarily kept distinct, but only to discourage speculation. It was not narrow-mindedness, he protested, but tactical advantage.[8] By the mid 1920s, however, Morgan and others were saying that the future of genetics lay in its connections with development and evolution—putting the gene back in the whole organism, as some liked to put it. And it was not just talk: Drosophilists were beginning to invest in these new lines. In 1923–24 Charles Metz tried to entice Sewall Wright to Cold

7. Garland E. Allen, *Thomas Hunt Morgan: The Man and His Science* (Princeton: Princeton University Press, 1978), pp. 301–17.

8. T. H. Morgan, "The theory of the gene," *Amer. Nat.* 51 (1917): 513–44, on pp. 535–36.

Spring Harbor to lead a program in physiological and developmental genetics.[9] By the late 1920s Morgan was telling his Carnegie patrons that development was where big new discoveries would be made.[10] By 1930 most of the fly group's new recruits were choosing dissertation topics in the genetics of development, not classical genetics. By the mid 1930s faith in developmental genetics was drosophilist gospel, and every major research group was competing to hire young people in these lines. In 1936 Milislav Demerec made it a priority to appoint someone to his staff in developmental genetics, and in 1939 he told CIW president Vannevar Bush that the future of genetics lay in development and evolution.[11] L. C. Dunn made developmental and evolutionary genetics the basis of his program to restock Columbia with drosophilists and put its genetics group back on the map.[12]

Despite their programmatic commitment, however, drosophilists had little practical success in reuniting genetics with development and evolution. They devised many ingenious experiments but none, before the mid 1930s, that were really productive of significant results. When Metz wanted to begin work in developmental genetics in 1923, he felt that too little had been done even to suggest what lines might be productive.[13] Morgan's 1926 lecture on development and genetics, though fervent about the possibilities, contained virtually no specific cases of how the two fields might be connected.[14] The situation had not improved much by the mid 1930s, when Morgan published his book on embryology and genetics. *And* was the crucial word: Boris Ephrussi recalled how, in the summer of 1934, he had read with eager anticipation

9. C. Metz to C. B. Davenport, 24 Dec. 1923, 29 Jan., 12, 19 Mar. 1924; [Metz], "Memorandum for Dr. Davenport," 25 Nov 1923; [Metz], "General statement of plans—1924"; all in CIW file Genetics.

10. Morgan to Merriam, 4 Jan. 1927, CIW; and Morgan to Max Mason, 13 May 1933, THM-CIT 1.

11. M. Demerec, "Memorandum on the gene problem," 24 Jan. 1936, MD file Gene Study; and Demerec to V. Bush, 8 Apr. 1939, CIW file Drosophila Stock Centers.

12. E. W. Sinnot and L. C. Dunn, "Memorandum for the president concerning genetics and development," in Dunn to F. D. Fackenthal, 12 Oct. 1939, LCD file Columbia University Genetics Lab; and Dunn to T. Dobzhansky, 24 Apr. 1938, TD.

13. [Metz], "Memorandum for Dr. Davenport," 25 Nov. 1923, CIW file Genetics.

14. T. H. Morgan, "Genetics and the physiology of development," *Amer. Nat.* 60 (1926): 489–515.

Morgan's advance copy of the book only to find that there was almost nothing in it about the genetics of development. When Ephrussi summoned his courage and suggested that the title was misleading, Morgan replied that it was not in the least misleading, that it did in fact contain "some embryology and some genetics."[15]

But if it was not lack of interest that impeded efforts to reconnect genetics with development, then what was it? The evidence suggests that it was mainly the difficulty of inventing experimental systems that could compete with mainstream practices.

Making Do with *Drosophila*

In principle drosophilists could simply have teamed up with embryologists to study the genetics of development (as Bcadle and Ephrussi did in the mid 1930s). That, however, was easier said than done in the 1910s and 1920s. The trouble was that the practices of these two communities had evolved along divergent paths, with different preferred organisms. It is indicative of the gap between them that the basic embryology of *Drosophila* was not done until the late 1920s and early 1930s, thus precluding any direct study of genes that controlled morphogenesis. It was not just the inherent difficulty of the embryology that put people off, but the social division of labor: the drosophilists' work offered few incentives to acquire the demanding skills of experimental embryology, and experimental morphologists, who had those skills, naturally preferred to exercise them on more amenable material than tiny flies. As Jack Schultz observed, both parties had chosen their experimental materials for their particular convenience, thus ensuring that they would have little to say to each other.[16]

Geneticists and embryologists took an interest in each other's work mainly in connection with control experiments that were designed to keep their modes of practice pure and separate. Embryologists had to be aware of genetic effects so that they could hold them constant (e.g., by using only homozygotes or inbred strains) and thus avoid mistaking genetic for developmental phenomena. Drosophilists likewise had to pay attention to physiological and developmental effects, which might

15. B. Ephrussi, "The cytoplasm and somatic cell variation," *J. Cellular Compar. Physiol.* 52 suppl. 1 (1958): 35–53, on p. 36; and T. H. Morgan, *Embryology and Genetics* (New York: Columbia University Press, 1934).

16. J. Schultz, "Aspects of the relation between genes and development in *Drosophila*," *Amer. Nat.* 69 (1935): 30–54, on p. 52.

mimic genetic phenomena and lead the unwary into embarrassing blunders of interpretion. It is striking how many of the examples that Morgan cited in 1926 of transdisciplinary experiments were in fact control experiments—preserving distinctions, keeping distance.[17] This kind of transdisciplinary interaction did little to unite the two camps, since it was meant to sharpen and protect boundaries and not to find common ground. It was the production process itself that prevented rapprochement: the credibility of both parties depended on their ability to keep genetic and developmental phenomena separate. What was signal to one group was, to the other, noise.

Drosophilists who were interested in doing developmental genetics generally tried to improvise new uses for their familiar, proven genetic systems. It was not an unreasonable thing to do. *Drosophila* was by its nature a dual-purpose instrument: genes were known only by their visible effects on morphological characters. Drosophilists assumed that a chain of developmental reactions intervened between the primary reactions of the genes and the visible morphology—how long a chain no one knew. However, one could in principle illuminate the early steps of the developmental process by studying how different combinations of genes influenced characters. Practically, that meant co-opting for developmental purposes experimental systems originally designed for doing chromosomal mechanics. But the trouble with these improvised systems was that experiments designed to illuminate development somehow kept turning into experiments on segregation and crossing-over. It proved difficult to co-opt for other purposes a system of production that had built into it the ambitions and skills of a whole generation of genetic mappers. Drosophilists modified their standard organism in some highly ingenious ways, but try as they might, their improvisations produced mainly knowledge of chromosomal mechanics.

A number of purely genetic phenomena seemed especially well suited to studying the genetics of development, especially genic interactions. The phenomenon that seemed most promising in the 1920s and 1930s was the so-called position effect, in which the expression of a gene is altered by other, unrelated genes located nearby it on the chromosome. This phenomenon was first studied by Sturtevant in 1925 in the bar-eyed mutant and was opened up for systematic exploration after 1928 by the abundance of large deletions, inversions, duplications, and translocations that H. J. Muller found were produced by X-raying. Drosophilists reasoned that position effects would lead them to the mystery of gene action, since interactions between adjacent

17. Morgan, "Genetics and the physiology of development."

genes must reflect the very earliest of the chain of reactions that connected genes to physical traits.[18] It was logical to think so, given current notions of how genes worked, but the results were nonetheless discouraging. As Curt Stern admitted in 1939: "The position effect . . . leads us to the door behind which primary gene actions occur; however, the door still remains sealed."[19] A good deal was learned about genic interactions, but little about development.

Drosophilists also had high hopes for studies of gene dosage, that is, the effects on characters of different doses of a gene in haploids, diploids, or triploids. New trick systems developed in the 1920s and 1930s opened up abundant opportunities for virtuoso experiments on dosage effects: for example, Bridges's high nondisjunction mutant, which threw off large numbers of haploids and triploids, or Muller's deletion and duplication mutants. No less promising were the large and ever-growing families of developmentally related mutants, especially the eye-color series, in which many instances were known of genes modifying the effects of others. These effects, it was hoped, might offer clues to the sequence of developmental reactions and intermediates between genes and characters. As Schultz put it: "The behavior of genes in combination with each other forms a bridge between the description of development in the different mutant races—what may be called Mendelian embryology—and the study of gene action proper." Again, however, the results were disappointing. Schultz had to admit that no one had been able to figure out a way of combining studies of gene dosage with real embryology.[20] Similarly, work on pleiotropy (the ability of one gene to affect many characters) was no more illuminating of the process of development, nor were other genetic systems in which genes modified the effects of other genes in the formation of characters.

We can see in these varied studies of position effect, gene dosage, and modifier effects how the dual potential of *Drosophila*, plus the social imperatives of production and publication, operated to cause deviant practices to revert to the mainstream. Having invested in genetic skills and lacking comparable skills in experimental morphology, drosophilists naturally began by modifying genetic systems, which they knew how to use and which they knew worked. When these systems did not pro-

18. Schultz, "Aspects of the relation," p. 51; Stern, "Report on the grant of $5400," [1941], CS file RF; and Carlson, *The Gene*, chaps. 13–15.

19. C. Stern, "Recent work on the relation between genes and developmental processes," *Growth* 3 suppl. (1939): 19–36, on p. 25.

20. Schultz, "Aspects of the relation," pp. 42 (quotation), 45–51.

duce the dramatic insights into development that had been expected, drosophilists naturally focused on whatever genetic knowledge they did produce. Improvised systems had this genetic fallback built into them, and that helps explain why tinkering with standard genetic systems so seldom resulted in really novel systems for doing developmental genetics. The systems used to study dosage and position effects were designed for exploring the intricacies of crossing-over, and they produced such knowledge even in experiments that were meant to do other things. Genetic agendas were likewise built into drosophilists' expectations, ambitions, and career calculations. Careers in experimental science are not made by reporting failures, after all, but by making the most of what is produced. These built-in capacities and investments set practical limits on the ability of drosophilists to redesign a mapping instrument as a tool of developmental genetics.

It was the same on the physiological and morphological side of *Drosophila* work, but in reverse. Here, rather than working from genes to characters, experimenters began with characters, altering their development by manipulating the environment and then attempting to correlate patterns with genetic constitution. In these "phenogenetic" studies, as they were known at the time, drosophilists (and others) improvised upon standard physiological methods to make them more genetic. Drosophilists interested in the physiology of development usually studied the effects of temperature, nutrition, and humidity on the development of normal or mutant characters. Such experiments were variants of the experiments that Morgan's more biological students had done in the 1910s on the physiology of crossing-over. The difference was that in the 1920s they were aimed at understanding development rather than genetic transmission. It was doubtless because this mode of experiment was so familiar and assessable that it was so widely practiced.

A good example of this type of experiment is the work of Charles Plunkett on the development of bristles in *Dichaete* and other mutants.[21] Though not important morphologically to the fly, bristles were very important for the fly people because they varied markedly with genetic and environmental conditions and could be counted and measured quantitatively. Similar studies were done with the family of *scute* bristle mutants by George P. Child, of *vestigial* wing by Morris H. Harnly, and of *bent* wing by Charles Metz and other alumni of the Columbia fly group. Jack Schultz exposed larvae to brief pulses of heat at various

21. C. R. Plunkett, "The interaction of genetic and environmental factors in development," *J. Exp. Zool.* 46 (1926): 181–244.

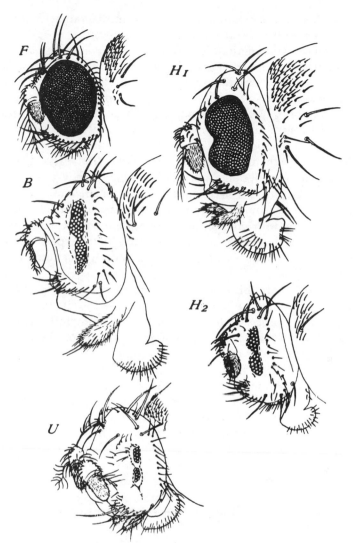

Full-eye, bar, ultra-bar, and their heterozygotes. *F*, full-eye wild females. *B*, white-bar female of the second generation of low selection from which ultra-bar was derived by mutation (culture no. 144.2). *U*, ultra-bar female (no. 158.1). *H₁*, heterozygous female from full × bar. (no. 87.11). *H₂*, heterozygous female from full × ultra-bar (no. 259.3). × 60.

Figure 6.1   *Bar* eye series of mutants, used in experiments on the genetics of development by Charles Zeleny and his students. From Charles Zeleny, "A change in the *bar* gene of *Drosophila melanogaster* involving further decrease in facet number and increase in dominance," *J. Exp. Zool.* 30 (1920): 246.

points in their development in the hope of finding out at what stage in the developmental process aberrations occurred.[22]

This type of experiment was pursued most systematically by Charles Zeleny and his students at Illinois, most notably Joseph Krafka, Jr., who studied the variation of eye-facet development with temperature in the *Bar* series of mutants. This program in developmental genetics evolved out of control experiments that Zeleny had performed in connection with his earlier evolutionary studies of variation and selection. At first, temperature effects were a nuisance in genetic studies, which Zeleny sought to understand and control. In the 1920s, however, when development became more fashionable, Zeleny began to treat temperature effects as a significant problem in their own right. Like bristles, eye facets would be counted and treated statistically, and they varied markedly with environmental conditions, especially temperature.[23] Once an annoying complication in genetic experiments, variability of development became a strategic site for a genetics of development.

As it turned out, this type of experiment produced a great deal of data on the physiology of development but shed no light on the role of genes in the process. Morphological experiments produced morphological knowledge, just as genetic experiments produced genetic knowledge. The major conclusion from Plunkett's and Krafka's work was that development behaved kinetically as if it were a chemical reaction, which was old news. Phenogenetic studies, like position and gene dosage effects, did not open a window on the developmental process, because they were not sufficiently different from standard practices. These modes of work were attractive because they were familiar and offered the prospect of innovation without requiring major investments in difficult new skills. But for the same reason such improvisations were unlikely to produce modes of experiment that truly integrated genetics and development. Similar patterns can be seen in the work of experimental zoologists who used organisms other than *Dro-*

22. Schultz, "Aspects of the relation," pp. 31–33.

23. C. Zeleny, "A change in the bar gene of *Drosophila melanogaster* involving further decrease in facet number and increase in dominance," *J. Exp. Zool.* 30 (1920): 292–324; Zeleny, "The effect of selection for eye facet number in the white bar-eye race of *Drosophila*," *Genetics* 7 (1922): 1–115; J. Krafka, Jr., "Development of the compound eye of *Drosophila melanogaster* and its bar-eyed mutant," *Biol. Bull.* 47 (1924): 143–49; and Krafka, "Environmental factors other than temperature affecting facet number in the bar-eyed mutant of *Drosophila*," *J. Gen. Physiol.* 3 (1920–21): 207–10.

sophila. Their numerous studies were pure experimental morphology; genetics entered in, if at all, only in the initial construction of stocks.[24] Just as studies of gene dosage and position effect were reassimilated into mainstream genetics, so, too, did "phenogenetic" experiments tend to become mainstream experimental morphology. It was relatively rare for anyone to invest equally in genetic and developmental methods. Krafka did some work on the embryology of eye facets, but he was exceptional. Most geneticists and developmental biologists simply did what they knew how to do and called it developmental genetics.

Developmental Genetics: Sturtevant

One might have expected that Morgan, with his active programmatic interest in development and evolution, would take the lead in devising new experimental methods. He did not, however, at least not with *Drosophila.* Characteristically, he seems to have rested his hopes on finding new standard organisms. In the late 1910s, for example, he was deeply involved in a study of sexual selection and the development of secondary sexual characters in Sebright fowl. In the early 1920s he published the results of similar experiments on feathering in birds and the asymmetric claws of fiddler crabs.[25] Was he, one wonders, trying out new lines of research on the genetics of development and searching for the right organism, as he had before with mice, *Phylloxera,* and *Drosophila?* Was he hoping to recapitulate in developmental or evolutionary genetics the earlier success with *Drosophila* but with new organisms? One can only surmise, but such an interpretation is consistent with Morgan's familiar working habits: for each line of experiment its particular organism. In any case, it was Sturtevant and Schultz who were boldest and most active in inventing new methods of developmental and evolutionary genetics. They were doubtless encouraged by Morgan's example and programmatic statements, but unlike Morgan they sought to realize their ambitions using *Drosophila.*

Sturtevant took an active interest in development as early as the late 1910s.[26] His most significant contribution occurred in work on

24. Valentin Haecker, *Entwicklungsgeschichtliche Eigenschaftsanalyse (Phaenogenetik),* (Stuttgart: Fischer, 1918); and Haecker, "Phänogenetisch gerichtete Bestrebungen in Amerika," *Z. induk. Abstam.* 41 (1926): 232–38.

25. A. H. Sturtevant, "Thomas Hunt Morgan," *Biog. Mem. Nat. Acad. Sci.* 33 (1959): 283–325; and Morgan to A. F. Blakeslee, 19 Oct., 23 Nov. 1916, AFB.

26. E. B. Lewis, "Alfred H. Sturtevant," *DSB* 13: 133–38.

gynandromorphs and mosaics. These were female flies in which some cell lines lost an X chromosome during development, resulting in patches of tissue or whole organs or body sections that were genetically different from the tissues of the rest of the mosaic fly (figure 6.2) The earlier in the process of development the X chromosome was lost, the larger the mosaic parts of the adult fly. These genetic curiosities had been eagerly collected whenever they happened to turn up, because they were potentially useful to Bridges in his work on sex determination.[27] Sturtevant, however, was more intrigued by their potential as tools for studying development.

Because mosaics arose at different stages in embryogenesis, they were potentially an instrument for dissecting the process of development itself, which purely genetic and phenogenetic methods never could. First, mosaic patches could provide clues to the point in embryogenesis when, say, bristles or eye facets differentiated. In principle it was even possible to construct entire cell lineages from the early germ layers to the adult organs. Such genetic dissection was an attractive alternative to physical dissection, which was very hard to do with dipteran flies.

A second type of experiment with mosaics gave another tantalizing glimpse of the physiological mechanism of development. The trick here was to study how the development of a mosaic patch was affected by surrounding tissue, the genetic constitution of which could be manipulated at will by inserting mutant genes into the X chromosome. In such experiments mosaics served as naturally occurring genetic analogues of embryologists' transplantation experiments, in which bits of tissue were relocated physically to different positions in the embryo to see if their development was altered. Generally, development was "autonomous," that is, characters developed in mosaic patches independently of surrounding tissues. In a few cases, however, development was nonautonomous, presumably because it had been affected by chemical substances diffusing into the mosaic patch from surrounding tissues. Such cases were treasured because they offered an opportunity to observe the physiological links in the chain of reactions between gene and character.

Sturtevant found two such cases of nonautonomous development, one in 1920 and one about 1926. The first was a mosaic eye that was genetically *vermilion* but developed as wild type, owing to the diffusion

27. T. H. Morgan and C. B. Bridges, "The origin of gynandromorphs," in *Contributions to the Genetics of Drosophila melanogaster,* CIW publication no. 278 (Washington, 1919), pp. 1–122.

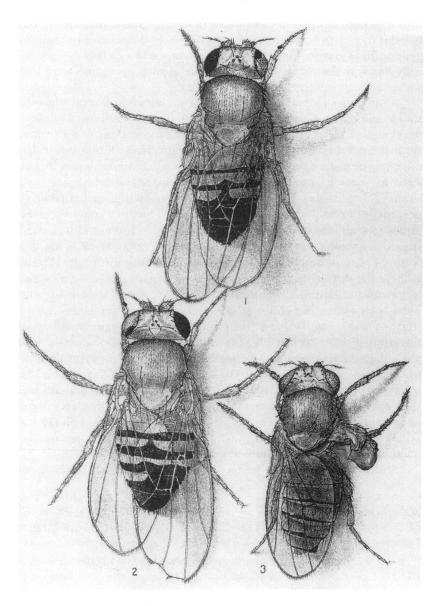

Figure 6.2   Gyanandromorph mutants, used by Alfred Sturtevant in experiments on development of eye color and other characters. Sturtevant, *Origins of Gynandromorphs,* plate 4, at p. 54.

into it of pigment precursors that were lacking in the mutant eye but present in nearby wild-type tissue. In the second case an eye that was genetically wild type showed some bar character in a fly that was genetically *Bar*, and vice versa; again, diffusable growth hormones were presumed to be involved.[28]

As an experimental system, however, mosaics had several practical shortcomings. They appeared rarely, unpredictably, and usually late in development, producing only small patches and making it difficult to study the developmentally crucial early stages. These limitations ruled out any systematic study. The prospects of a systematic program brightened in 1926, however, when Sturtevant discovered a peculiar *claret* strain of *D. simulans* that threw off an unusually high proportion (3 percent) of early-cleavage gynandromorphs. This trick system seemed finally to open the door to a real genetics of development. Sturtevant began a systematic study of cell lineage. When a *vermilion* mutant turned up, he seized the chance to find out where in the developing embryo the diffusable pigment precursor was made. He also found two more cases of partial nonautonomous development (*scute* bristle and *yellow* body color) and collected *Minute-n* mosaics in preparation for a study of bristle embryology. Sturtevant co-opted younger members of the group into his project. He persuaded Theodosius Dobzhansky, who was a skilled morphologist, to dissect developing mosaics and find out when mosaic effects appeared. He set a visitor from China, Ju-Chi Li, the task of finding out at what stage of development lethal genes had their effect.[29] Sturtevant was not farming out marginal problems to unimportant visitors. Rather, he was making use of visitors' morphological skills—essential skills, which the fly group did not have—to build a major program in developmental genetics.

Sturtevant's new trick systems did not, however, pay off as richly as

28. A. H. Sturtevant, "The vermillion gene and gynandromorphism," *Proc. Soc. Exp. Biol. Med.* 17 (1920): 70–71; Sturtevant, "The effects of the bar gene of *Drosophila* in mosaic eyes," *J. Exp. Zool.* 46 (1927): 493–98; and Metz to Demerec, 24 Feb. 1923, MD.

29. A. H. Sturtevant, "The use of mosaics in the study of the developmental effects of genes," *Proc. VI Int. Cong. Genetics (1932)*, 1: 304–307; Sturtevant, "The claret mutant type of *Drosophila simulans:* A study of chromosome elimination and of cell-lineage," *Z. wiss. Zool.* 135 (1929): 323–56; Sturtevant, *Contributions to the Genetics of Drosophila simulans and Drosophila melanogaster,"* CIW publication no. 399 (Washington, 1929), pp. 16–17, 51–52; Sturtevant to Stern, 8 Jan. 1927, CS; Dobzhansky to Stern, 2 Nov. 1937, CS; Bridges to Stern, 1 Feb. 1928, CS; and Ju-Chi Li, "The effect of chromosome aberrations on development in *Drosophila melanogaster,"* Genetics 12 (1927): 1–58.

he had hoped. The cell-lineage work proved exceedingly tedious and boring, and there were too few cases of nonautonomous development to make the mosaic method generally useful. Mosaics are not true developmental mutants, that is, mutants of genes that control morphogenesis. They are the result of mutations in genes that regulate the orderly segregation of chromosomes during cell division (nondisjunction), which just happen to act during development rather than gametogenesis. Although they brought drosophilists closer to development than any other trick systems, mosaics were essentially variants of the standard fly, which produced knowledge of nondisjunction and sex determination—for which purposes they were originally saved. The real importance of Sturtevant's work, in hindsight, lay less in its results than the challenge it set to younger and bolder members of the fly group, like Jack Schultz and George Beadle, to find methods that were both genetic and developmental.

Developmental Genetics: Jack Schultz

Of all the fly group, Jack Schultz was the one who invested the most in problems of gene action and development. He was greatly influenced by Sturtevant, "even to the point of overstimulation," as he recalled.[30] No one, not even Sturtevant, was more ingenious in devising developmental uses for triploids, piebald mosaics, deficiencies, and other variants; Schultz was a virtuoso improviser. Yet he published very little of this work, in part because he disliked writing, but also because most of his projects never fulfilled the grand hopes he had for them. His ideas were bold and brilliant, but they did not translate into productive experimental systems. Schultz was pushing traditional genetic systems to their limits, ingeniously exploiting the developmental potential of *Drosophila* but always being drawn back to alternatives in chromosomal mechanics. Developmental variations could not compete in practice with the rich rewards of mainstream "Drosophilistics."

Schultz's hallmark style is apparent in his dissertation work on the *Minute* series of bristle mutants. The question Schultz asked was, How many genes and how many distinct reactions are involved in the development of bristles? Schultz hoped to find the answer by constructing various combinations of the twelve *Minute* genes, two at a time, and seeing which combinations showed a dosage effect. If two different *Minute* genes reinforced each other, he reasoned, then they must control

30. Schultz to Beadle, 31 July 1970, JS.

the same developmental reaction.[31] In theory, *Minute* mutants should form clusters of genes that interacted with each other but not with genes in other clusters. Each cluster would represent one reaction in the developmental chain. It was a compelling idea, but unfortunately none of the forty-nine different combinations that Schultz constructed displayed a dosage effect. Schultz tried some other tricks to salvage the project, but was very glad to leave the *Minutes* behind.[32]

In the early 1930s Schultz systematically combed other families of mutants for evidence of developmental clustering, especially the large family of twenty-five eye-color mutants. Ingeniously, Schultz looked not to the adult character of these mutants, but rather to their secondary, developmental characteristics: for example, early or late appearance of color, or side effects on bristle color. By sorting mutants into groups according to their developmental characteristics, Schultz hoped to correlate patterns of genic interactions with reactions in the developmental chain. In fact, the eye mutants did indeed seem to form two or possibly three clusters, and Schultz proceeded to synthesize stocks with various combinations of genes from the same and from different groups, to see if genes modified each other only when they were in the same cluster group. In over one hundred double-mutant stocks he did in fact find nine cases of modifying effects but, alas, no correlations with his clusters.[33] Schultz's labors were rewarded with nothing more than tantalizing hints.

Schultz also spent a great deal of effort in the late 1920s on a bold project that combined genetics and biophysics in the study of eye pigments. The idea here was to probe a single developmental process as it was going on. If methods could be devised for assaying eye pigments and their chemical precursors, the process of their development could be followed quantitatively. Schultz hoped in this way to dissect the chain of chemical and developmental reactions by which pigments

31. Since *Minute* mutations were lethal in homozygotes, Schultz assumed that a double dose of different *Minute* genes would also be lethal if they affected the same developmental reaction. The lethal effect would make these combinations easy to recognize.

32. J. Schultz, "The minute reaction in the development of *Drosophila melanogaster,*" *Genetics* 14 (1929): 366–419; Schultz to Stern, 26 Feb.–4 Mar. 1926 [sic: 1927], 9 Sept. 1928, CS; Helen Redfield to Stern, 14 June 1927; CS; and Schultz to Morgan, 3 Jan. 1929, JS.

33. Schultz, "Aspects of the relation," pp. 39–43, 47; and Schultz to Stern, 16 Oct. 1929, 12 Feb. 1930, CS.

were synthesized.[34] In 1926 Schultz won a National Research Fellowship to study with Selig Hecht, a biophysicist who specialized in color vision and retinal pigments. As it happened, Hecht had just become a member of Columbia's department of zoology, thanks to Morgan, who lobbied his fellow zoologists and got a foundation grant to outfit Hecht's laboratory.[35]

Schultz spent two years with Hecht learning to separate and measure different eye pigments by their optical and chemical behavior. "The necessary thing for the kind of work in which I am really interested is a process in development which can be singled out *during development*," he wrote Stern. "The eye colors are the nearest to this that I know: so for the last year I have been grinding heaps of Drosophilas in mortars, extracting pigment, and performing various kinds of tests."[36] With a quantitative assay he hoped to be able to measure the kinetics of pigment development in flies with various combinations of eye-color genes, and from patterns of genic interactions infer what genes controlled what steps in pigment synthesis. Genes that controlled the same reactions would modify each other, he reasoned, while those that controlled different reactions would not. It was like the *Minute* experiments but better, because the rate of formation of pigment was a developmental characteristic, not an end product: "Once given the technique, all of the experiments that seemed so unsatisfactory to me with the Minutes, become possible in elegant terms."[37]

Such work was unlike anything drosophilists were accustomed to doing, and perhaps for that very reason Schultz seemed to relish it.

34. Schultz, "Aspects of the relation," pp. 33–39; Schultz to Morgan, 3 Jan 1929, JS; and Schultz, "Progress report of work done during tenure of National Research Council Fellowship, 1928–29," n.d. [1929], 13, JS. Eye-color mutants seemed ideal for such a project because a great deal was known about their histology and development.

35. Schultz, "Proposed studies of the physiology of development," n.d. [1926 or 1927], JS; Morgan to H. J. Thorkelson, 2 Feb. 1926, GEB 1.4 681.7040; S. Hecht to W. J. Crozier, 11, 29 Mar. 1926, William J. Crozier Papers, Pusey Library, Harvard University; and George Wald, "Selig Hecht (1892–1947)," *J. Gen. Physiol.* 32 (1948): 1–16, on p. 5.

36. Schultz to Stern, 9 Sept. 1928, CS.

37. Ibid.; Schultz to Stern, 16 Oct. 1929, CS; Schultz, "Report of work done during tenure of a National Research Council Fellowship," n.d. [1928], Schultz, "Progress report of work done" [1929], pp. 11–13; Schultz, "Final progress report of work done during tenure of National Research Council Fellowship, 1928–29," n.d. [1929], and Schultz to Morgan, 3 Jan. 1929, all in JS.

Figure 6.3 Jack Schultz at Caltech in the early 1930s, about the time that George Beadle and Boris Ephrussi were there. Courtesy of Jill Schultz Frisch.

When the fly group left for California in 1928, Schultz stayed behind to continue his work with Hecht. (Morgan offered Hecht a position at Caltech but was turned down.) Schultz would have gone to Germany in 1929 to study pigment chemistry with Otto Warburg and Richard Willstätter, but Morgan made it plain that he expected him to take up his new job as Bridges's successor.[38]

As it turned out, the pigment work was not much more productive than Schultz's other endeavors. The work got off to a slow start at Pasadena, when Schultz had to wait the better part of a year for Morgan to supply him with a spectrophotometer.[39] Meanwhile, he became interested in more orthodox genetic problems, and the pigments languished. Characteristically, Schultz simply could not stick to one line of work. His friends tried to keep him on the straight and narrow path, but to little avail. "Schultz does many things, including again some eye pigments," Dobzhansky wrote Stern in 1933. "I told him that he should do pigments and only pigments, and offered to him to go in his room and to inject carbon tetrachloride in all his cultures with the exception of those containing eye color genes. So far he has not granted permis-

38. Schultz to Stern, 9 Sept. 1928, CS; Schrader to Stern, 7 June 1928, CS; and Schultz to Morgan, 28 Jan. 1929, JS.

39. Schultz to Stern, 12 Feb., 2 Aug. 1930, CS.

Figure 6.4 Helen Red-
field Schultz at Columbia,
circa 1925–28. Courtesy of
Jill Schultz Frisch.

sion to do this drastic operation, but I am glad to say that he seems very
much inclined to grant it soon."[40] Schultz's interest in development did
revive in 1934–35, when he published his seminal paper on genes and
development. But developmental genetics remained just one among
his varied genetic interests, and in 1937 he abandoned eye pigments
for good and went to Sweden to work with Torbjörn Caspersson on
the biochemistry and biophysics of chromosomes. That seemed a more
direct and promising approach to the chemical structure and function
of genes.

It was not Schultz's restlessness alone that killed his work on development, but also the inherent limitations of his experimental system.
The eye-color mutants were essentially a genetic system, with a vast in-
ventory of genetic experience built into and around it. Knowledge of
embryogenesis had to be inferred indirectly from genetic data. So
when experiments designed to study development did not prove fruitful, it was easy to carry on with the purely genetic aspects. Schultz
learned a good deal about genic interactions, but not much about development. No one came closer to giving *Drosophila* a developmental
capability; no one was as bold and imaginative in devising hybrid experiments with methods borrowed from other disciplines. But his work on

40. Dobzhansky to Stern, 1 Nov. 1933, CS.

Figure 6.5   Jack Schultz and Torbjörn Caspersson, Stockholm, 1938.
Courtesy of Jill Schultz Frisch.

Minute and eye-color pigments reveals the limits of what it was possible
to do in development with genetic systems of production.

In retrospect Schultz's greatest accomplishment was to show
how the social boundary between genetics and other disciplines could
be crossed. Geographically it was not far from Schermerhorn Hall to
Hecht's biophysics lab—Schultz had only to cross the Columbia campus and ascend to the thirteenth floor of the new physics building.
In the psychosocial topography of disciplines and practices, however,
Schultz bridged a gulf when he decided to put two such different
modes of practice together, and his example deeply influenced
younger members of the fly group.

Evolutionary Genetics: Charles Metz

We see the same pattern of improvisation and retreat in drosophilists'
efforts to unite genetics and evolution. Here, too, we see much ingenuity and effort expended in refitting *Drosophila*, with little to show at
the end: lots of genetics, but none of the hoped-for insights into the
evolutionary process. Wild species, brought into the laboratory with

the aim of making them into specific instruments for evolutionary genetics, ended up being rebuilt to the standard design of *D. melanogaster* and assimilated into mainstream genetic practice. Drosophilists tried to use standard flies in new ways to study phylogenetic relations, but the work proved extremely laborious and unproductive, and experimenters either gave up or reverted to classical chromosomal mechanics.

Wild drosophilas were brought indoors in the hope that they would produce interspecies hybrids. In theory such hybrids would have offered a simple and direct way of determining phylogenetic relationships, because chromosomes that were homologous in different species (because they had evolved from a common ancestor) would visibly pair. Unfortunately, species did not form fertile hybrids (the few exceptions were of limited use because their chromosomes were so similar). Consequently, drosophilists were forced to fall back on tried-and-true methods: namely, establishing homologies by systematically mapping the chromosomes of the different species and looking for similar sequences of genes. This fallback method was not only exceedingly slow and laborious, it also had the effect of making wild species into variants of the standard laboratory fly, because standardization was a necessary preliminary to comparative mapping. The result of comparative mapping was not a new system for evolutionary genetics, as drosophilists hoped, but an enlargement of the standard system of chromosomal mechanics. Thus were the first efforts to reunite genetics and evolution pulled back to the mainstream.

Charles Metz was probably the first to bring new wild species of *Drosophila* into the lab, beginning soon after he arrived at Columbia in 1912. By October 1913 he was already deep into his dissertation work with E. B. Wilson on the comparative cytology of the Diptera. Metz was always more interested in comparative cytology than in genetics per se, and at first *Drosophila* occupied a minor place in his plans. The dipteran flies had an evil reputation among cytologists for being difficult to work with. They rose in Metz's esteem, however, when he found that with the proper fixing and staining techniques they gave chromosome preparations that were quite clear and easy to read. Indeed, the chromosome figures of *Drosophila* were especially striking because homologous chromosomes tended to stay paired together more than in most species. What Metz saw through his microscope looked like the idealized artist's rendering of paired chromosomes in textbook illustrations![41]

41. Metz to Davenport, 31 Oct., 23 Dec. 1913, 27 Apr., 7 May 1914, CBD; and C. Metz, "Chromosome studies in the Diptera, II: The paired association

Metz realized that *Drosophila* was in fact an ideal creature for doing comparative cytology and phylogenetics. It had relatively few chromosomes (four to six), and different species had distinctly different numbers and shapes of chromosomes: large V's, short rods, hooks, and dots. The more species Metz put under his microscope, the more types of chromosomal groups he found and the more hopeful he became that they could be arranged in a sequence that corresponded to the order in which species had evolved by splitting up of large V-shaped chromosomes into two short rod-shaped ones, or fusing of two into one, or losing or gaining small dot-shaped chromosomes. So excited was Metz by this prospect that he boldly put forward (in his very first publication) a phylogenetic tree of the twelve known chromosome group types (figure 6.6).[42]

Morphological resemblances alone were a treacherous guide to phylogenetic relations, however, as Metz soon learned. Because chromosomes come in just a few shapes and sizes, it is risky to assume that similarity of shape is evidence of a genetic relationship. In fact, the more types of chromosome groups turned up, the more arbitrary all phylogenetic schemes seemed to be. Repenting of his haste, Metz became extremely cautious about drawing any conclusions about homologies and phylogeny from cytological evidence alone.[43]

There were only two ways to be certain that chromosomes were homologous. One could map them and see if they had the same genes in the same order: that was the hard way. The easy way, at least in theory, was to construct a hybrid of the two species and see what chromosomes paired together. For example, if a large V-shaped chromosome of one species had split during speciation, then it should pair with two

---

of chromosomes in the Diptera, and its significance," *J. Exp. Zool.* 21 (1916): 213–79.

42. C. Metz, "Chromosome studies in the Diptera, I: A preliminary survey of five different types of chromosome groups in the genus *Drosophila*," *J. Exp. Zool.* 17 (1914): 45–49, on p. 48; Metz to Davenport, 31 Oct., 2, 23 Dec. 1913, 1 Apr. 1914, CBD; and Metz, "Report for the year ending September 1st 1915," CBD.

43. Metz, "Report for the year ending September 1st 1915," CBD; C. Metz, "Chromosome studies on the Diptera, III: Additional types of chromosome groups in the Drosophilidae," *Amer. Nat.* 50 (1916): 587–99; and Metz and M. Moses, "Chromosome relationships and genetic behavior in the genus *Drosophila*," *J. Hered.* 14 (1923): 195–204, on pp. 200–202.

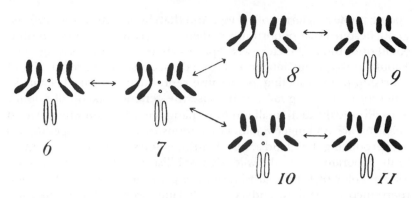

Figure 6.6 Phylogenetic relations between different *Drosophila* species, inferred by Charles Metz from chromosomal figures. From Charles Metz, "Chromosome studies in the Diptera, I: A preliminary survey of five different types of chromosome groups in the genus *Drosophila*," *J. Exp. Zool.* 17 (1914): 48.

rod-shaped chromosomes in a hybrid with its daughter species.[44] Easy, that is, if drosophilas of different species would produce viable hybrids.

Genetic mapping was at first a minor part of Metz's program. Metz was aware what a coup it would be to extend the idea of linkage groups to other species of *Drosophila*. Nevertheless, mapping remained an emergency backup in case the search for hybrids failed.[45] Metz was repelled by the practical difficulties of systematic chromosomal mapping. It was hard, slow work and not suited to a small laboratory like his. Mapping a dozen species would have been a gargantuan task, and Metz kept putting it off. In 1915 he had fifteen to twenty species of *Drosophila* in cultivation and was hunting mutants in four. But even the most promising species had thus far produced only four or five mutants suitable for mapping. Metz put his efforts mainly into the search for interspecies hybrids.

For six or seven years Metz labored to construct a variety of hybrids, but all his diligence and ingenuity were to no avail. Interspecies hybrids were in fact not like the trick system of *D. melanogaster*. They could not be constructed by simply applying known tricks of the trade. Drosophilas of different species had to cooperate, by courting, mating, and pro-

44. Metz, "Brief statement of scientific program for Drosophila investigations," n.d. [Jan 1919], CBD file Drosophila Investigations.

45. Metz to Davenport, 2 Dec. 1913, 2 Apr. 1914, CBD; and Metz, "Report for the year ending September 1st 1915," CBD.

ducing viable and fertile offspring. And that, by and large, they refused to do, despite every opportunity and encouragement that drosophilists could offer them. It is clear why. Species are formed when distinct sub-populations can no longer interbreed, because divergent habits of courtship prevent mating, or because reproductive organs became so different that mating fails, or because the genetic makeup of populations diverge through small random changes to the point where hybrid offspring are no longer viable. Populations that have no impediments to reproducing their kind are by definition one species. It was the same in the laboratory as in the wild. Drosophilists could manipulate genes for eye color or bristles at will, but the genes for courtship behavior, reproductive anatomy, and genetic balance were virtually unknown and far too complex to be redesigned and reengineered. When it came to species hybrids, the fly people were dealing with an aspect of wild nature that could not be reconstructed for life in the lab.

Drosophilists could always hope, of course, that odd species or local subspecies existed in obscure corners of the natural world that would form hybrids. The odds of finding them were slim; however, the payoff of finding species that would hybridize would have been so great that drosophilists continued the search even in the face of continual disappointments. At first Metz and Sturtevant collected near home, around New York City, at Woods Hole, up and down the Atlantic coast. Having about exhausted the varieties that they could collect easily in the northeastern United States, they then began to cast their eyes further afield, to the tropical regions where drosophilas flourished on the abundance of rotting fruits and vegetation. Metz was again first to go afield in a big way. In fall 1914 the head of Cold Spring Harbor, Charles Davenport, agreed to underwrite a collecting expedition in the Caribbean and Central America, and in January 1915 Metz departed for Cuba.[46]

Metz was astonished by the rich pickings within just a few miles of his base near Havana:

> I thought I had seen a good many Drosophilas up around New York and Cold Spring Harbor, but I have changed my mind during the last few days. There are more Drosophilas within 100 yards of where I am sitting than in the whole state of new York I am quite sure. At any rate I can go out here and catch more in an hour or two than I could in weeks at home. It is simply astounding! . . . I think it is safe to say that we have

46. Metz to Davenport, 31 Oct. 1913, 25 Sept. 1914, CBD; and Metz, "Report of cytological and genetic work done during the months June to September, inclusive, 1914," CBD.

already taken as many species here as we have obtained in the whole United States, or very nearly that many at least. And only two or three of them are species found in the U. S.[47]

Metz was also impressed by how much easier it was to culture drosophilas in a tropical climate and hoped that his new captives would be more inclined to make hybrids if bred there on the diet and in the climate to which they had been accustomed. (The Metzes had also, like their flies, become fond of the climate and the abundant tropical fruits.) However, Davenport vetoed a field station in Cuba.[48]

Metz returned to New York with twenty-five species that could be bred in the lab, including a dozen that were fecund enough for genetic work—but still no interspecies hybrids. He switched back and forth between trying laboratory tricks and hoping for a stroke of luck with wild flies. If the drosophilas did not interbreed naturally, he would try to devise a technique of artificial insemination, and if that did not work he would go back to the field. "I am confident that hybrids can be secured," he assured Davenport. "It only depends upon getting the right species, and since the world is full of Drosophilas, especially the tropical and sub-tropical regions, I am a long way from giving up hopes of securing suitable ones." He was especially hopeful about two regional varieties of *D. repleta*, which had different chromosome types but might still be sexually compatable.[49]

Hybrids and Evolutionary Genetics: Sturtevant

Metz had by this time (1915) been joined by Sturtevant in the search for hybrids. Of all the fly group Sturtevant was always the keenest natural historian and taxonomist. In 1911 he spent the summer at Cold Spring Harbor working on sexual selection, and in 1913 he opened a notebook on the mating habits of *Drosophila*.[50] He collected bugs when-

47. Metz to Davenport, 30 Jan. 1915, CBD.

48. Ibid.; Metz to Davenport, 1, 24 Mar., 9 Apr. 1915; and Metz, "Report for the year ending September 1st 1915"; all in CBD.

49. Metz to Davenport, 9 Apr. 1915; see also Metz to Davenport, 24 Mar. 1915; and Metz, "Report for the year ending September 1st 1915," pp. 6–7; all in CBD.

50. A. H. Sturtevant, "Experiments on sex recognition and the problem of sexual selection in *Drosophila*," *J. Animal Behav.* 5 (1915): 351–66; and Sturtevant, "Notes on the mating habits of the Diptera,"AHS 16.

ever he found himself in the countryside: in the suburban woods of Fort Lee, just across the Hudson River, at Woods Hole, with his wife's family in upstate New York, and at his own family's farm in Kushla, Alabama. (The Gulf Coast offered especially rich and varied microecologies.) He was working on a systematic taxonomy of Drosophilidae as early as 1916.[51]

It appears that Sturtevant first took a systematic interest in *Drosophila* hybrids about 1914, as a result of participating in Metz's work on comparative cytology. In 1914–15 Metz and Sturtevant collaborated in hybridization experiments with all the species that they had in hand.[52] Their complementary skills, Metz's in cytology and Sturtevant's in taxonomy and systematics, made such a collaboration a natural one. Sturtevant also participated in the quest for more exotic species. While Metz was in Cuba, Sturtevant was in Panama and Central America on his own collecting expedition, and on their return to New York in 1915 they again collaborated in an effort to make hybrids from the new tropical species that they had brought back.[53] All in all they did about 150–200 different crosses. The most promising species (most similar in appearance and mating rituals or most frequently observed courting) were investigated "very extensively."[54] Although Sturtevant never again mounted a full-scale field expedition, the search for wild species continued in more modest ways. For example, it became a custom for members of the fly group to collect wild species whenever they traveled to visit families or went on vacation. Morgan took the group to Califor-

51. A. H. Sturtevant, "Notes on North American Drosophilidae with descriptions of twenty-three new species," *Annals Entom. Soc. Amer.* 9 (1916): 323–43; Sturtevant, "Acalypterae (Diptera) collected in Mobile County, Alabama," *J. N.Y. Entom. Soc.* 26 (1918): 34–40; Sturtevant, *The North American Species of Drosophila* (CIW publication no. Washington, 1921); and Sturtevant to Demerec, 10, 24 July 1926, MD.

52. Morgan to Woodward, 13 Feb. 1920, CIW; and C. Metz, M. S. Moses, and E. D. Mason, *Genetic Studies on Drosophila virilis with Considerations on the Genetics of Other Species of Drosophila,* CIW publication no. 328 (Washington, 1923), pp. 7–8.

53. Sturtevant, "Chronology," n.d., AHS 4.2; and Morgan to Payne, 29 Sept., 28 Nov. 1915, FP.

54. T. H. Morgan, C. B. Bridges, and A. H. Sturtevant, *The Genetics of Drosophila* (The Hague: Nihoff, 1925), p. 184; Morgan, "Program of work," to Woodward, 2 Oct. 1917, CIW; and Sturtevant to Mohr, 11 Aug. 1919, 10, 25 Sept. 1919, OM.

nia in 1920–21 in part to get access to the *Drosophila* species of the Pacific slope, and there was talk of collecting in Hawaii.[55]

Although little came of all this effort, the search for species hybrids was taken very seriously in New York and Cold Spring Harbor in the late 1910s. Despite repeated failures, a lucky find was always possible and would have been an experimental gold mine. No wonder Morgan reacted so violently to Metz's proposal in 1919 that the two Carnegie groups divide up *Drosophila* species between the two places.[56]

Several cases of interspecies hybrids did in fact turn up, but they proved disappointing. In 1919 Sturtevant identified a hybrid between *D. melanogaster* and a newly identified species, *D. simulans*, that turned up unexpectedly in cultures of standard *melanogaster* stocks. Unfortunately, the hybrid flies were sterile and thus useless for genetic work. (The sterility of offspring in certain crosses was, in fact, what tipped Sturtevant off that an unidentified species had been living unnoticed in what seemed to be homogeneous cultures of *D. melanogaster.*) Nor were these hybrids of use for comparative cytology, since *melanogaster* and *simulans* had identical chromosome shapes.[57]

Rather more promising was the hybrid that was turned up in 1922 by Sturtevant's student, Donald Lancefield, in *D. pseudoobscura.* Lancefield had been constructing genetic maps of this species when he stumbled by accident on two stocks from widely separated regions of the United States, which bred with difficulty.[58] This hybrid was at the time

55. Morgan to Woodward, 13 Feb., 25 Sept. 1920, CIW; and Allen, *Morgan,* pp. 289–94.

56. Morgan to Metz, 26 Feb. 1919, THM-APS.

57. Sturtevant to Mohr, 2 Dec. 1919, OM. Hybrids were useful, however, for checking whether similar mutations in the two species were allelomorphic: Sturtevant to Mohr, 9 July 1921, 20 Feb. 1922, OM. *Simulans* flies look like small, pale *melanogasters* (they are distinguishable only by the different genitalia of the males). Because drosophilists customarily selected large, dark flies for experimental crosses, they almost never deliberately mated *melanogaster* with *simulans,* and since the offspring of such a mating would have been sterile, any such experiment would have been discarded as a failed experiment. *Simulans* survived, however, because it had a higher tolerance than *melanogaster* did for old, dried-up food. A. H. Sturtevant, "Genetic studies on *Drosophila simulans,*" *Genetics* 5 (1920): 488–500; 6 (1921): 43–64, 179–207; and Sturtevant, *Contributions to the Genetics of Drosophila simulans and Drosophila melanogaster,* CIW publication no. 399, (Washington, 1920), pp. 1–62.

58. Morgan to R. Goldschmidt, 15 Feb. 1921, RG; and Sturtevant to Mohr, 29 Jan., 4 Mar., 6 Apr., 9 July 1921, 3 Dec. 1922, 17 Mar. [1923], OM.

not thought to be a hybrid between two species, since the parental stock, named A and B, were believed to be geographical races that were in the process of speciation. Although female A-B hybrids were sufficiently fertile for genetic analysis, they were not much use for phylogenetic study, since their chromosome types were identical.[59] It was a dilemma: the more two species differed, the more interesting their hybrids were as experimental tools, but the more impossible it became to construct them. *Pseudoobscura* found some special use in investigating the genetic cause of hybrid sterility, but it did not appear to be the long-sought key to an evolutionary genetics.[60]

By the late 1920s drosophilists had ceased to hope that hybrids offered an easy way to an evolutionary genetics. Metz had by then given up *Drosophila*, though Sturtevant continued to hope that artificial insemination might succeed where nature failed—Alfred Huettner was working on that at New York University.[61] In fact, more systematic collecting would eventually turn up numerous cases of interspecies hybrids, no fewer than eighty-six by 1950, some of which produced fertile offspring. Interestingly, no new hybrids turned up among the *melanogaster* group of species, nor did they among any cosmpolitan species. Hybrids, it turned out, are formed most easily between species that have evolved by adapting to specialized diets in similar microhabitats—fungus and pollen feeders, and the like. But these species are also the hardest for laboratory workers to find and to cultivate in the lab.[62] None of this was clear in the 1920s, of course, and in the absence of species hybrids, evolutionary genetics was gradually absorbed into the genetic mainstream.

59. Donald E. Lancefield, "A genetic study of two races or physiological species of *Drosophila obscura*," *Z. induk. Abstam.* 52 (1929): 287–317. First thought to be an American form of the European *D. obscura*, this species was renamed *pseudoobscura* in 1929. Race B was reclassified as a distinct species, *D. persimilis,* in the 1940s: Lancefield to Stern, 1 Nov. 1930, CS; and William B. Provine, "Origins of the genetics of natural populations series," in *Dobzhansky's Genetics of Natural Populations, I–XLIII,* ed. Richard C. Lewontin, J. A. Moore, W. B. Provine, and B. Ames (New York: Columbia University Press, 1981), pp. 5–79, on pp. 21–26.

60. Lancefield to Stern, 1 Nov. 1930, CS; Sturtevant to Stern, 8 Jan. 1927, CS; Sturtevant to Clausen, 27 Aug. 1927, CS; Lancefield to Dunn, 6 Apr., 5 May 1929, LCD; and H. Redfield to Muller, 13 Mar, 1930, HJM-C.

61. Metz to Demerec, 8 July 1929, MD; and Morgan to Merriam, 14 Dec. 1928, CIW.

62. John T. Patterson and W. S. Stone, *Evolution in the Genus Drosophila* (New York: Macmillan, 1952), pp. 384–85, 393, 423.

Comparative Genetics

As their hopes for fertile hybrids waned, drosophilists had no alternative but to determine homologies the hard way, by genetic mapping. Metz and Sturtevant mapped several species in the early 1920s, and Lancefield began to map *D. pseudoobscura*. However, the main result of their labors was to turn these wild species into minor variants of the standard fly. The new species ended up becoming second-string instruments of mainline genetics, not as versatile and productive as *melanogaster*, and valued mainly for their capacity for special little tricks. *D. virilis*, for example, had an unusually high rate of mutation.

This process of assimilation can be seen most clearly in the changing character of Metz's work after his return from war work in 1919. Mapping, once an emergency backup, became the central activity of his expanded program, thanks to the new resources that he got as a result of the fracas with Morgan. Metz set out to map *D. virilis* and *D. willistoni*, the best of his wild species, just as Bridges had mapped *D. melanogaster.* He constructed standard stocks and tamed the natural variability of his wild flies by inbreeding. Scaled-up crossing experiments produced numerous new mutants.[63] *Virilis* and *willistoni* thus joined *melanogaster* as domesticated breeder reactors. Metz was not interested in mapping for its own sake but only as a means of revealing phylogenies. However, his genetic means gradually became an end in itself as mapping brought him no closer to his evolutionary goal.

At first Metz found the epidemic of mutants as exciting as the fly group had in the early years, but the excitement soon wore off. It had all been done before. For Metz, mutant hunting and mapping were necessary but boring routine, which were best delegated to technicians. It was not something that he liked to do himself, and gradually his interests turned elsewhere.[64] With his new junior colleague, Milislav Demerec, he tried to create new chromosomal types artificially, using physical methods such as centrifugation to disrupt the machinery of crossing-over. And, increasingly, he turned to other insects, which he

63. Metz, "Plans for the coming year," n.d. [1918?], CBD; and Metz to Davenport, 14 Nov., 8 Dec. 1918, 27 Jan., 22 Feb., 13, 18 Apr. 1919, CBD.

64. Metz to Davenport, 21 Aug. 1923 and "1923 Report", CBD; and [Metz], "Memorandum for Dr. Davenport," 25 Nov. 1923, CIW file genetics. Difficulties also arose in using genetic maps to determine chromosomal homologies, since similar-looking mutants in different species could not be assumed to be the same: Rebecca Lancefield and Charles Metz, "The sex-linked group of mutant characters in *Drosophila willistoni*," *Amer. Nat.* 56 (1922): 211–41.

Figure 6.7   Charles Metz,
circa 1920s. Courtesy of
American Philosophical Society Library, Stern Papers,
photographs file Metz.

hoped might be more amenable than drosophilas to crossing species lines. One insect in particular caught his fancy, the fungus gnat *Sciara*, and by 1926 Metz was working with *Sciara* exclusively and with the same enthusiasm that he had once felt for *Drosophila*.[65]

Metz later told friends that he had stopped working on *Drosophila* because of pressure from Morgan.[66] This seems unlikely, given Metz's stubbornly independent character: no one pushed him around. It is much more likely that Metz gave up on *Drosophila* because he was discouraged by his failure to find hybrids and was bored with genetic mapping. He took up with *Sciara* because he thought it would produce interspecies hybrids and thus be a better tool than *Drosophila* for what he wanted to do.[67] Genetics was for Metz never an end in itself but a tool of experimental evolution. When mapping did not bring him closer to his goal, he had no reason to continue working with *Drosophila*.

Applying standard genetic methods to wild species of *Drosophila* did not create a new mode of evolutionary genetics. On the contrary, the practice of comparative genetic mapping transformed wild dro-

65. Metz to Davenport, 12, 19 Mar. 1924, and Metz, "General statement of plans—1924," CIW file Genetics; and Metz to Davenport, 30 June, 7 July 1924, 11 July, 125 Aug. 1926, 28 June, 5 Dec. 1927, all in CBD. *Sciara* had the odd habit of congregating in the dark interiors of parked automobiles, which thus became the world's biggest and most expensive gnat traps.

66. Demerec to Gilbert, 10 Nov. 1939, CIW file Bridges.

67. Metz, "General statement of plans—1924," CIW file Genetics.

sophilas into standard laboratory flies. The result of comparative genetic mapping was not phylogenetic trees but genetic maps, and the nascent field of evolutionary genetics was thus reabsorbed into mainstream practice. Drosophilists continued to map wild species, but those who did so in the 1920s and early 1930s, such as Warren Spencer, H. Kikkawa, and D. Moriwaki, used wild species as variants of *D. melanogaster*, confirming and recapitulating what had been established before with the standard fly.

It was the practical realities of production, I believe, that shaped drosophilists' choices. The search for hybrids among exotic wild species promised great rewards, but it was all or nothing: if you did not find hybrids, your whole investment of time and labor was lost. Comparative mapping was laborious and old-fashioned, but it was more forgiving. Should comparative mapping fail to produce insights into phylogenies, drosophilists were still left with their maps, which were valued by their fellow workers, and with new standard flies, which might prove useful for some special study of chromosomal mechanics. This fallback position was built into drosophilists' system of production and made it easy for them to revert to mainstream chromosomal mechanics.

Conclusion

In the long view, pure chromosomal mechanics dominated *Drosophila* genetics for a surprisingly short period of time. Drosophilists were trying to develop a comparative, evolutionary genetics as early as 1914, a year before the first full chromosome maps were published, and serious efforts in developmental genetics were under way by the mid 1920s. A decade later these two fields were seen by ambitious young drosophilists as the places where reputations were likely to be made in future. The new systems of the mid 1930s—Beadle and Ephrussi's transplantation method, and Dobzhansky's genetics of natural populations—may seem to have appeared suddenly from nowhere, but in fact they grew out of the improvisations of Metz, Sturtevant, Schultz, and others that went back almost twenty years.

It is remarkable how long drosophilists had to tinker with existing systems before hitting upon really productive ways of doing developmental and evolutionary genetics. Their efforts were varied and ingenious, yet most were reabsorbed into mainstream practice. Work at the research front took the shape of a series of forays, retreats, and more forays. This pattern may be fundamental to all experimental work, and its causes seem to lie in the nature of the production process itself, in the imperatives of getting return from investments in craft skills and ensuring continuity of production when systems run dry. It is the practicalities of working and making careers that explain why drosophilists preferred to contrive new uses for standard experimental systems rather than to seek radically new ones outside the mainstream.

First, the drosophilists' social customs and work culture encouraged a strategy of improvisation. They had become accustomed to a fast-paced, highly productive mode of work that discouraged radical retooling. In an experimental culture like the fly group's it would have been unusual and discomforting to be unproductive for very long; self-esteem would suffer, even if careers did not. We would expect such a culture to encourage these habits of improvisation and opportunism. We would expect individuals who worked in such a culture to be aggressive in inventing new uses for their experimental systems, ever hopeful of finding some striking new things. We would also expect them not to pursue unproductive lines too long, but to do what was most doable and wait for fresh opportunities to reopen difficult problems. Hence the pattern of repeated initiatives and retreats that we see so strikingly illustrated in the drosophilists' work on evolutionary and developmental problems in the 1910s and 1920s.

Second, the material culture of *Drosophila* genetics encouraged this same strategy of improvisation and reconnoitering. The most productive trick systems of the 1920s, like triploids, duplications, and mosaics, had been devised for work on chromosomal mechanics, but they were also potentially useful in studying development. Conversely, exotic wild species of *Drosophila* were brought into genetics laboratories to be used in studying phylogenies, but they were potentially also "standard" flies, especially when hybrids were abandoned for comparative mapping. The inherent capacity of these experimental systems to generate problems in chromosomal mechanics gave drosophilists a fallback position: When their improvisations in evolution and development failed to live up to expectations, it was easier for experimenters to return to mainstream problems than to push on to more radical, riskier approaches.

In principle, drosophilists could have abandoned *Drosophila* for creatures more suited to experiments on development or evolution, or invested in novel experimental methods from other disciplines. And indeed they occasionally did: Sturtevant worked extensively with *Oenothera* in the late 1920s. But in general the Columbia drosophilists remained loyal to the fly, in a way quite unlike Morgan's (or Metz's) mobile partnerships with different creatures. The drosophilists' enormous investment in the standard fly—their creation—inclined them to tinker rather than to seek radically new approaches. With their unequaled craft knowledge of *Drosophila* and skill in improvisation, what incentives did they have to give up their competitive advantage for untried methods borrowed from other disciplines? Sturtevant knew in his mind that physiology and biochemistry were likely to be important for the future of genetics, but his heart was in classical chromosomal mechanics.[68]

68. Sturtevant to Mohr, 20 Oct. 1932, 19 Dec. 1926, OM.

SEVEN

Reconstructing *Drosophila*:
Developmental Genetics

I N THE 1940S AND 1950S developmental and evolutionary genetics
became the vanguard of *Drosophila* work and put traditional lines
of chromosomal mechanics in the shade. These new modes of
experimental life began in the mid to late 1930s, when drosophilists
invented new experimental systems that were as productive as the standard fly had been in mapping and chromosomal mechanics twenty-
five years before. On the developmental side the key event was George
Beadle and Boris Ephrussi's invention in 1935 of a method of transplanting imaginal disks in *Drosophila*. (Imaginal disks are the buds of
tissue in pupae from which adult characters develop.) Less dramatic
but no less pregnant for future work were the methods invented by
Ernst Hadorn, Donald Poulson, and others for studying the embryology of developmental mutants. Equally striking changes occurred on
the evolutionary side when Theodosius Dobzhansky in the United
States and others in the Soviet Union invented methods of studying
genetic change in local populations of drosophilas in the wild. Two
decades earlier, Morgan, Frank Lutz, and others had envisioned a kind
of experimental evolution that united laboratory and field practice; in
the 1930s their hopes were realized, though in ways they could hardly
have foreseen.

The novelty of these experimental modes of practice lay in their
being transdisciplinary hybrids, which adapted techniques from other
disciplines to *Drosophila*. In the case of developmental genetics, techniques were imported from experimental embryology and biochemistry—transplantation, transfusion, nutritional bioassay. On the evolutionary side, classical genetic practices were hybridized with methods
from entomology and field biology and, in the 1940s, with theoretical
biometrics and demography. It was this mixing of different material
cultures and practices that distinguishes the events of the late 1930s
from the earlier improvisations of Metz, Sturtevant, Schultz and others.
With the notable exception of Schultz's foray into biophysics, few drosophilists had foraged for experimental methods outside the boundaries of their own community of practice. Beadle, Ephrussi, Dobzhan-

sky and others did, and as a result dramatically extended *Drosophila*'s experimental range—and drosophilists' as well.

The question is, What made these young drosophilists more adventuresome than their elders? And is it a coincidence that the two most fruitful of the new innovations in practice were made at Caltech? Or was there something in this particular local context that made the barriers between domains of practice easier to surmount? Morgan, of course, generally encouraged foraging in embryology and evolution. Moreover, Sturtevant and Schultz offered in their own work concrete examples for aspiring boundary crossers. I hope to show, in fact, that Beadle and Ephrussi were inspired to invent a hybrid experimental mode by specific features of Sturtevant's work on mosaics and of Schultz's work on eye pigments. Dobzhansky's invention of the genetics of natural populations grew out of collaborations with Schultz and Sturtevant. Beadle and Ephrussi's invention of transplantation in *Drosophila* grew out of a collaboration that was stimulated and made possible by Morgan, Sturtevant, and Schultz. These collaborative efforts were local events and took their character from a specific, local context of practice; however, they also exemplified more general trends in the drosophilists' working world.

The experimental culture of the fly group, with its traditions of collaborative work and assimilating newcomers, created situations that encouraged the invention of hybrid experimental modes. Collaborative research brought people together in active and intimate relationships that were focused on specific practical tasks. And when these people happened to have different outlooks and skills, the potential for hybridization was high. The greater size and diversity of the community of drosophilists in the 1920s and 1930s, and the greater mobility made possible by foundation support of postgraduate research fellowships, made it more likely that visitors to the fly group would in fact be of different backgrounds. The fly group's customs, which in the 1910s created a common culture and identity among its members, in the different circumstances of the late 1920s and 1930s encouraged boundary crossing and hybrid practices.

It is difficult to get an accurate count of the number of visitors to the fly group, but there is no question that the traffic in postgraduate and other visitors was especially intense in the 1930s, more than it had ever been before. Despite competition from other centers, the Caltech fly group was still the drosophilists' main communication center and their favored resort for training. The intensity of collaboration in the fly group is more easily measured, and it was remarkably high between 1927 and 1940 (table 7.1). Over half of Jack Schultz's publications

Table 7.1
Coauthorship in the Caltech Fly Group, 1927–1940

	Sturtevant	Bridges	Dobzhansky	Schultz	Other	Percent
Sturtevant	16	0	8	1	6	48
Bridges	0	21	2	1	12	41
Dobzhansky	8	2	43	5	10	36
Schultz	1	1	5	8	3	56

SOURCES: Bibliographies in Francisco J. Ayala, "Theodosius Dobzhansky," *Biog. Mem. Nat. Acad. Sci.* 55 (1985): 163–228; Morgan, "Calvin Blackman Bridges," ibid., 22 (1941): 31–48; Thomas F. Anderson, "Jack Schultz," ibid., 47 (1975): 393–423; A. H. Sturtevant, "Thomas Hunt Morgan," ibid., 33 (1959): 283–325; and *Genetics and Evolution: Selected Papers of Alfred H. Sturtevant*, ed. E. B. Lewis (San Francisco: Freeman, 1961), pp. 320–26. I excluded from the data annual reports to the CIW, reviews, abstracts, and general papers.

were coauthored, and about two-fifths of Bridges's and Dobzhansky's!

Thus a local custom of collaboration and a demographic trend toward greater diversity and mobility together produced a situation that encouraged transdisciplinary foraging and increased the likelihood that diverse material cultures would form a successful hybrid.

The actual experience of change, however, was anything but straightforward. Beadle and Ephrussi, in showing how embryological transplantation could be applied to *Drosophila*, initiated a process of change that went well beyond their original expectations; indeed, it led Beadle eventually to adopt a new organism, *Neurospora*, and a mode of biochemical work that he himself did not find entirely congenial. At first, experimental practices were guided by a programmatic vision—the desire to invent a practicable system for developmental genetics. After that, however, one new twist led to another, causing Beadle's conception of developmental genetics to change radically. Transplantation led Beadle and Ephrussi to biochemical problems and to methods borrowed from nutritional biochemists. Those methods in turn revealed the shortcomings of *Drosophila* as an instrument of chemical genetics and impelled Beadle to discover a creature better suited to that work. The hybrid character of transplantation created the potential for these far-reaching changes in practice; however, in the actual process of change that potential unfolded gradually—epigenetically, so to speak. The process of change was experienced as a succession of different experimental systems in which fundamental changes emerged unexpected and unbidden from apparently quite small and innocent changes in experimental procedure.

The transformation of standard *Drosophila* practices into new modes of developmental and biochemical genetics was driven mainly, I will argue, by the imperatives of the work process itself—the imperatives of getting the most out of an experimental system, maintaining credibility among fellow experimentalists, and staying ahead of potential competitors. In the specific context in which Beadle and Ephrussi (and Dobzhansky) worked, these practical imperatives were powerful incentives to exploit novel experimental systems opportunistically, whatever one's original intentions. The nature of their experimental way of life gave drosophilists the flexibility to allow one thing to lead to another.

There are striking parallels between Beadle and Ephrussi's experience in the 1930s and Morgan's in 1910–12. Brought indoors to do experimental evolution, *Drosophila* first revealed its special potential for neo-Mendelian experimental heredity and displaced mice, then made it impossible to do neo-Mendelian experiments in the usual way and drove Morgan and his "boys" to invent a more doable kind of experimental heredity. A similar succession of experimental modes unfolded in Beadle and Ephrussi's work, as the potentialities of transplantation and nutritional biochemistry were revealed. As Morgan had been led to a mode of experimental genetics that he did not find entirely congenial, so, too, was Beadle drawn into a mode of biochemical genetics that he did not particularly enjoy. They were impelled to change by the productive power of their experimental systems and by the benefits of these systems for their young collaborators, for whom stripped-down and fast-paced experimental modes were entirely congenial ways of life.

To understand how chromosomal mechanics was reunited with development, we need to follow *Drosophila* through its microevolutionary succession, from standard fly, to instrument of biochemical genetics, and finally to its displacement by *Neurospora* from its honored place in Beadle's laboratory.

Boundary Crossers

Beadle and Ephrussi were both hesitant immigrants from other disciplines. George W. Beadle was a maize geneticist, a farm boy from Wahoo, Nebraska, who had been drawn gradually into a scientific career. He took his graduate training (Ph.D. 1931) at Cornell with Rollins A. Emerson, whose group occupied a place in the world of maize genetics that corresponded to Morgan's among drosophilists. Beadle was bright, ambitious, and competitive, but also extremely personable and diplomatic, a natural leader. (He succeeded Morgan at Caltech in 1945 and went on to become president of the University of Chicago.) Until his arrival at Caltech in 1931 he worked exclusively on classical chromosomal mechanics, using "sticky chromosome" mutants of maize (mutants in which chromosomes did not separate normally after synapsis). Beadle did not come to Caltech to do developmental genetics; in fact, he was not even sure that he wanted to be a drosophilist. When he won a National Research Council fellowship he meant to stay on at Cornell and continue his work on maize, but the NRC's fellowship board insisted that he broaden his horizons by working on a different organism.[1]

Relations between Emerson's maize group and the drosophilists were quite close at the time. Maize geneticists were involved in a big communal mapping project and so had good reason to cross over, so to speak, with drosophilists. Milislav Demerec, who got his Ph.D. with Emerson in 1923, worked on both organisms and thought all geneticists would profit from using *Drosophila* in addition to their own special systems.[2] So, too, did Ernest Anderson, another Emerson student and future member of the Caltech group.[3]

Nevertheless, boundary crossing was no easy feat. Beadle recalled how bewildered he was at first by the drosophilists' arcane jargon, and he hedged his bets by continuing his work on maize with his friend Sterling Emerson (Rollins's son). When Sterling began to teach himself *Drosophila*, Beadle joined him in a collaborative study of the mechanics of the attached-X system—the closest thing there was in *Drosophila* to Beadle's "sticky chromosome" system in maize.[4] Beadle thus charted the easiest route from maize to *Drosophila* and took care not to burn bridges behind him. A remark to Demerec in 1932 suggests that he fully intended to return to maize genetics when his fellowship was over.[5]

Boris Ephrussi's background and aspirations were entirely differ-

1. Norman H. Horowitz, "George Wells Beadle," *Biog. Mem. Nat. Acad. Sci.* 59 (1990): 27–52; G. W. Beadle, "Recollections," *Ann. Rev. Biochem.* 43 (1974): 1–13, on pp. 5–6; and Beadle, foreword to A. H. Sturtevant, *Genetics and Evolution,* ed. E. B. Lewis (San Francisco: Freeman, 1961), pp. iii–iv.

2. M. Demerec to E. D. Dale, 17 Oct. 1929, MD.

3. R. A. Emerson to C. Metz, 17 Mar. 1919, CBD; E. G. Anderson to Demerec, 4 May 1924, MD; and Demerec to R. A. Emerson, 4 June 1924, 23 Apr. 1926, MD.

4. G. W. Beadle and S. Emerson, "Further crossing over in attached-X chromosomes of *Drosophila melanogaster,*" *Genetics* 20 (1935): 192–208.

5. Beadle to Demerec, 9 Jan. 1932, MD.

Figure 7.1 George W. Beadle with friend, Nebraska, probably early 1920s. Courtesy of California Institute of Technology Archives, Photo File, 10.24.

ent from Beadle's. Russian-born and Paris-trained, Ephrussi came to Caltech on a Rockefeller Foundation fellowship in 1934 with a track record of work in experimental embryology and a strong desire to unite development with genetics. Trained in the classical embryology of the sea urchin egg, Ephrussi became interested in using mammalian systems while working in Emmanuel Fauré-Fremiet's tissue culture laboratory at the Rothschild Institute for Physico-Chemical Biology. Ephrussi began to work on tissue culture around 1928, and after completing his Ph.D. in 1932, he began a new line on the genetics of development in a mutant strain of mouse that carried a lethal gene, hoping to discover when and where the mutant gene acted in the developmental process.[6] Never having been made to learn genetics—it was not much taught in France—Ephrussi applied for a fellowship to study in the United States. At first he planned to go to Columbia to work with

6. Herschel Roman, "Boris Ephrussi," *Ann. Rev. Genetics* 14 (1980): 447–50; Richard M. Burian, Jean Gayon, and Doris Zallen, "The singular fate of genetics in the history of French biology, 1900–1940," *J. Hist. Biol.* 21 (1988): 357–402; and B. Ephrussi, "Sur le facteur lethal des souris brachyures," *Compte Rendus Acad. Sci. Paris* 197 (1933): 96–98.

L. C. Dunn, who was a leading expert on the genetics and embryology of mice, but he was somehow deflected to Caltech and *Drosophila* genetics.[7]

Like Beadle, Ephrussi followed a path of least resistance at first, working on his lethal mice, which he had brought with him from Paris. His first project with *Drosophila*, suggested to him by Sturtevant, was on the embryology of the *scute-8* lethal mutant, a system obviously chosen because of its similarity to the lethal mice.[8] Recognizing Ephrussi's skills in experimental embryology, Sturtevant steered him to a problem that taught him basic *Drosophila* know-how while furthering the fly group's line of work in developmental genetics. Sturtevant was personally interested in *scute-8* because in mosaics patches of tissue carrying the lethal gene developed normally when surrounded by wild-type tissues. It seemed an ideal system for developmental genetics, and Ephrussi's skill in classical embryology would enable him to take it further than Sturtevant had been able to.

Thus Beadle and Ephrussi found themselves working side by side in a place where social mechanisms for boundary crossing were well developed. The question is, How did Beadle decide to drop chromosomal mechanics and join Ephrussi in a bold attempt to invent a new way of doing the genetics of development? Unfortunately there is little direct evidence, so we can only make plausible inferences from the circumstances in which the event occurred. Very probably the source of Beadle's inspiration was the fly group itself.[9]

Developmental genetics was a center of attention in the fly group just at the time that Beadle and Ephrussi were there together between January and July 1934. Everyone was interested in it. Sturtevant had

7. Ephrussi to Dunn, 12 Dec. 1932, 6, 20 Feb., 17 Sept. 1933, and Dunn to Ephrussi, 3 Jan. 1932 [*sic:* 1933]; LCD. J. B. S. Haldane, another mouse person, was also at Caltech in 1932 learning "Drosophilogy": Haldane to Dunn, 19 Oct. 1932, LCD.

8. Ephrussi to Dunn, 17 Sept. 1933, 21 Jan., 26 Feb., 1934, LCD; Dunn to Ephrussi, 30 Sept. 1933, 12 Mar. 1934, LCD; Sturtevant to Demerec, 21 Jan. [1934], MD; and Demerec to Sturtevant, 3 Feb. 1934, MD.

9. Lily Kay states that Beadle arrived at Caltech already captivated "by the biochemical puzzle of gene action, especially in relation to enzymology," but the only evidence for that view is Beadle's recollection that, when a graduate student, he had watched Rollins Emerson trying to persuade plant physiologists to use genetic methods: Lily E. Kay, "Selling pure science in wartime: The biochemical genetics of G. W. Beadle," *J. Hist. Biol.* 22 (1989): 73–101, on p. 77; and Beadle, "Recollections," p. 5.

chosen mosaics for the subject of his paper at the International Congress of Genetics in 1932. Schultz had revived his work on eye pigments in 1933, and in the spring of 1934 he was busy composing his review of work on genes and eye pigments for a lecture at Berkeley in June.[10] We may be sure, knowing Schultz, that his ideas were the subject of general conversation. Dobzhansky was also working on several projects involving the embryology of developmental mutants.[11] Ephrussi's work on lethals in mice was actively and critically discussed in a seminar soon after his arrival.

At the same time news was also filtering into the fly group that other groups were also trying to invent new methods of combining genetics and embryology. Late in the summer of 1934 Sturtevant received a letter from Curt Stern telling him that one of his co-workers (unnamed) had devised a method for transplanting imaginal discs in *Drosophila* pupae. This news caused some stir in the fly group. "The scheme of grafting imaginal disks sounds very exciting," Sturtevant wrote Stern. "If you can develop a grafting technique for Drosophila all this gynandromorph stuff will be out of date, and the attack on development can really go ahead. I certainly wish you luck—I can think of no technical advance that seems to me as desirable in Drosophilistics. Even the salivary chromosomes would be put in the shade."[12] Dobzhansky was no less excited when Sturtevant told him about it later that fall; he thought it "an advance of extraordinary importance."[13]

Was it this news from Stern that sparked Beadle and Ephrussi to try their own experiments with tissue culture and transplantation? Or Schultz's work on eye pigments, or Sturtevant's on mosaics, or Dobzhansky's on morphogenetic mutants, or all of the above? We will never know, but in any case the problem clearly seemed ripe for picking in these crucial months. Beadle and Ephrussi were inspired not by theo-

10. J. Schultz, "Aspects of the relation between genes and development," *Amer. Nat.* 69 (1935): 30–54.

11. Th. Dobzhansky and F. N. Duncan, "Genes that affect early developmental stages of *Drosophila melanogaster,*" *Arch. Entwick.* 130 (1933): 109–30; and Dobzhansky and C. B. Bridges, "The mutant 'proboscipedia' in *Drosophila melanogaster*—a case of hereditary homoosis," *Arch. Entwick.* 127 (1933): 575–90.

12. Sturtevant to Stern, 6 Sept. [1934], CS. This note is an addendum to a letter, dated 6 Sept., that can be dated by internal evidence to 1934. Stern's own letter was lost. The work by Stern and his co-worker appears not to have been published.

13. Dobzhansky to Stern, 14 Jan. 1935, CS.

Figure 7.2   Boris Ephrussi (left) and Leslie C. Dunn conversing, Cold Spring Harbor, 1941. Courtesy of American Philosophical Society Library, Dunn Papers, small photo album.

ries of genes and enzymes, which were as vague as they were ubiquitous, but by specific experimental efforts that were being made within the fly group and elsewhere to connect genetics and embryology. Here were real, practical incentives to try something bold.

It also seems that Beadle and Ephrussi were losing interest in what they were doing on attached-X chromosomes and lethal mice. Beadle later recalled that his work on crossing-over had not proved as fruitful as he had hoped in revealing the nature of the gene.[14] Ephrussi's discontent was more acute. Three months after returning to Paris, he wrote Schultz that he was unable to get back to business as usual. He was desperately "home seek [sic]" for the lively scientific life of the Morgan group, with its openness and freewheeling intellectual style. He had tried to recreate the group's seminar at the Rothschild Institute but had been obstructed by the Parisian scientists' refusal to take an interest in anything outside their own disciplines. He felt "asphyxiated." After Caltech he found it impossible to return to his research on experimental embryology. He was giving up work on lethal mutants in

14. Beadle, "Recollections," p. 6.

mice, his enthusiasm chilled by the sharp criticisms leveled at him by his Caltech friends, and he was at a loss to know what to do with tissue culture. What he really wanted to do, he knew, was "la vrai génétique," but his laboratory was not set up for that, and he did not know where to start—maybe chemical mutagenesis, he thought. Ephrussi was chagrined to recollect how he had failed to take advantage of friendly conversations with Morgan at Woods Hole in August to get some ideas for specific genetic projects. When would Beadle write again? Why did his other new friends not write? What was he missing? Had he been forgotten so quickly?[15]

Discontent with mainstream practices, plus Sturtevant, Schultz, and Stern's enthusiasm for developmental genetics, induced Beadle and Ephrussi to try that game themselves. So, with financial help from Morgan, it was arranged that Beadle would spend the first half of 1935 with Ephrussi at the Rothschild Institute in Paris and there try to apply methods of experimental embryology, such as tissue culture or transplantation, to *Drosophila*.[16] Thus a fragile bridge was thrown across the gap that separated embryology and genetics.

The collaboration of Beadle and Ephrussi was no coincidence. The fly group organized their collective work deliberately to encourage boundary crossing and new kinds of experimentation; that is why people like Beadle and Ephrussi went to them to learn genetics. Sturtevant had been working the edge habitat between genetics and embryology for a long time, and Schultz's foray into biophysics set an example for other would-be boundary crossers. Consciously or not, Beadle and Ephrussi were following Schultz's lead, but in a different and, as it turned out, far more fruitful way.

Inventing Transplantation

But how to adapt methods of experimental embryology to a creature that had been constructed to do pure genetics? Tissue culture was an obvious choice, given Ephrussi's expertise and the resources of the Tissue Culture Laboratory. Beadle later recalled that he and Ephrussi tried tissue culture first, failed, and only then turned to transplantation. But in a letter written one month after arriving in Paris, Beadle

15. Ephrussi to Schultz, 7 Nov. 1934, JS; see also B. Ephrussi, "The cytoplasm and somatic cell variation," *J. Cellular Compar. Physiol.* 52 suppl. 1 (1958): 35–53, on p. 36.

16. Beadle, "Recollections," pp. 6–7.

noted that they had not tried tissue culture and had begun directly with transplantation.[17] Most likely Ephrussi had tried tissue culture before Beadle arrived and concluded from his experience that transplantation would be a better bet.[18]

Transplantation was no sure bet, however. France's leading insect embryologist, Charles Perez, warned Ephrussi and Beadle that insect embryologists had always avoided flies because the imaginal disks all looked alike, making it almost impossible to pick out disks of specific organs, especially with such a tiny creature as *Drosophila*. Beadle and Ephrussi decided to try transplantation anyway. No other insect had anything like the rich accumulation of mutants and genetic potential of *Drosophila;* the potential reward on the genetic side therefore outweighed the risk on the embryological. Besides, Beadle was in Paris for six months whatever happened, so he had everything to gain and little to lose by taking a chance.

At first it seemed that Perez might be right. A month into the work Ephrussi and Beadle were just beginning to get the knack of picking out disks and injecting larvae without killing them, though still with no positive results. (Hoping to make life easier, Beadle asked Demerec to send a stock of the *giant larva* mutant.)[19] But one June morning—it must have been not long after Beadle wrote Demerec—he and Ephrussi came into the laboratory and found a newly hatched fly with a third eye clearly visible just beneath the skin of its abdomen. They knew then that they had unwittingly dissected out an eye disk. The question was, could they do it again on purpose? They retired to the Capoulade, a nearby coffee house, to plan their experimental campaign. It must have been obvious that the first thing to do was to test the *vermilion* mutant—in fact, Beadle had already asked Demerec to send him a culture of vermilion flies. It was the obvious thing to do because Sturtevant's work with vermilion mosaics led them to expect a positive result. Since the vermilion gene developed nonautonomously in mosaics, and since transplants were basically artificial

17. G. W. Beadle, "Genes and chemical reactions in *Neurospora*," *Science* 129 (1959): 1715–19, on p. 1716; Beadle, "Recollections," p. 6; and Beadle to Demerec, 5 June 1935, MD.

18. In late 1934 Ephrussi told Schultz of plans to investigate the nutritional requirements of "pure" tissues in culture; this may have been a preliminary to experiments with *Drosophila:* Ephrussi to Schultz, 7 Nov. 1934, JS. On Ephrussi's lab see Burian et al., "Singular fate," pp. 391–94.

19. Beadle to Demerec, 5 June 1935, MD.

mosaics, an eye disk from a vermilion larva, transplanted into the larva of a wild-type fly, should develop into a wild-type eye. And so it did.[20]

Beadle and Ephrussi knew at once that they had invented a general method of developmental genetics. Transplantation did everything that Sturtevant's mosaics could do, but with none of the limitations of a system that produced usable mosaics unpredictably and only for sex-linked genes. With transplantation, combinations of any genes could be tested by manipulating the genetic constitution of transplant and host.[21] Morgan quickly pronounced their work as important as Muller's invention of mutation by X-ray or Theophilus Painter's discovery of salivary cytology—no small compliment. Sturtevant eagerly spread the news and had Beadle send him a three-eyed fly to exhibit at a meeting of geneticists at Woods Hole.[22] Two modes of practice, previously distinct and carried out in different places, came together in one place, united in one hybrid practice.

The system that Beadle and Ephrussi had invented was an exacting bit of laboratory handicraft. Transplantation required close cooperation between two workers, seated opposite each other at a narrow table behind two binocular microscopes (see figure 7.3). One microscope was fitted with a dissecting stage and a microdissection apparatus to pick out imaginal disks; the other was fitted with a micropipette, to suck up the minute disk and inject it into the host larva. One person washed donor larvae with sterile water and dissected out the disks, while his partner washed the host, etherized it (not too fast, lest the larva shrink and curl up), and placed it in position for injection. The dissector then sucked the disk slowly into the micropipette and injected in to the side of the host larva, which his partner held in position with a blunt needle. This last and crucial operation required some little skill:

20. G. W. Beadle, *Genetics and Modern Biology* (Philadelphia: Saunders, 1963), pp. 11–12; Beadle, "Genes and chemical reactions," p. 1716; B. Ephrussi and Beadle, "A technique of transplantation for *Drosophila*," *Amer. Nat.* 70 (1936): 218–25; and Ephrussi to Beadle, 9 Sept. 1938, GWB 1.26.

21. B. Ephrussi and G. W. Beadle, "La transplantation des disques imaginaux chez la *Drosophile*," *Comptes Rendus Acad. Sci. Paris* 201 (1935): 98–100; and Beadle and Ephrussi, "Transplantation in *Drosophila*," *Proc. Nat. Acad. Sci.* 21 (1935): 642–46.

22. F. B. Hanson diary, 14–27 Aug. 1935, RF 1.1 205D 7.87; and Demerec to Beadle, 27 Sept. 1935, MD.

Chapter Seven

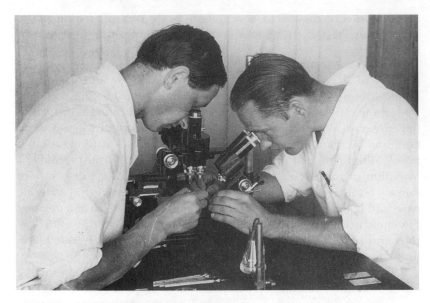

Figure 7.3 George Beadle and Boris Ephrussi performing a transplantation experiment on *Drosophila* larvae, mid-1930s. Courtesy of California Institute of Technology Archives, Beadle Papers, 2.2.

> The precise way of holding the host larva, the manner of inserting the pipette, the speed at which the tissue and saline should be injected and the rate and direction of withdrawal of the pipette and its relation to the rate of flow of liquid are all matters that can best be learned by experience. The most frequent difficulty and the hardest to learn to overcome is "blowing out" of the larva, *i.e.*, flowing out of part of the internal organs.[23]

With experience, however, it was possible to do thirty injections per hour and one to two hundred per day. On good days sixty percent of the larvae developed into adult flies, though the average was more like a third.

Determining the phenotype of the developed transplants also required craft skill. Transplanted eye disks, for example, usually developed deep within the abdomen, inside out (i.e., with the facets on the concave surface), and attached to genitalia, antennae and parts of the

23. Ephrussi and Beadle, "Technique of transplantation," p. 224; see also Ephrussi to Demerec, 26 Apr. 1937, MD.

head. It was no small task to get eyes out and cleaned up for inspection. An additional problem was that eye colors developed abnormally inside the host's body, so that elaborate controls were needed to make sure colors were identified correctly. (To determine a vermilion, or *v*, disk in a claret, or *ca*, host, for example, three control experiments were done in addition to the test, to provide reference colors.)[24] New routines for large-scale screening of transplants also had to be developed. Efficient methods for collecting large numbers of eggs had to be devised, as well as methods of culturing larvae with synchronized life cycles, so that hosts could be harvested at just the optimal age for transplantation.

Transplantation and Genetics

Transplantation did not immediately alter drosophilists' habits of using the fly as a genetic instrument. Most of the thirty-odd papers on transplantation that Beadle and Ephrussi published between 1935 and 1938 contained a great deal more genetics than embryology. The immense wealth of genetic knowledge and know-how that was built into the *Drosophila* system shaped Beadle and Ephrussi's experimental choices more effectively, it seems, than the embryological potential of transplantation. They spent most of their time at first exploring the interactions of eye-color genes, very much as Jack Schultz had, using genetic methods. Not surprisingly: transplantation made it much easier to study genetic interactions, since it eliminated the need to construct double-recessive test stocks. The only real problem that Beadle and Ephrussi encountered was getting a complete set of eye-color mutants, since there were none in France. "One has to work in a country where Drosophila doesn't exist," he observed, "to appreciate DIS and the cooperation that goes with it."[25] But with stocks rushed to them by Demerec, Beadle and Ephrussi worked their way systematically through the twenty-six known eye-color mutants, testing each in a wild-type host fly to see if eye color developed nonautonomously. These cases were the ones that would reveal where, when, and how genes were active in the biosynthesis of eye pigments. As it turned out, there were only two

24. G. W. Beadle and B. Ephrussi, "The differentiation of eye pigments in *Drosophila* as studied by transplantation," *Genetics* 21 (1936): 225–47, on pp. 226–27. The three controls were *v* disk in *v* host, *ca* disk in *ca* host, and *ca* disk in *v* host.

25. Beadle to Demerec, 13 Sept., 16 July 1935, MD.

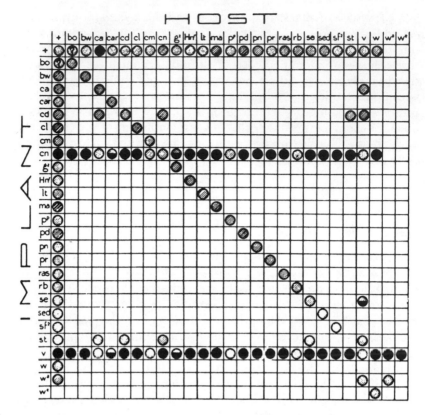

Diagrammatic representation of the results of eye transplants. Shaded circles indicate autonomous development of the pigmentation of the implant. Black circles indicate non-autonomous development of pigmentation. Circles half black and half shaded indicate non-autonomous development of such a nature that the resulting implant is intermediate in color between two controls.

Figure 7.4   Results of transplants of mutant eye disks in wild-type and mutant hosts. Black circles indicate nonautonomous development of eye color. From Beadle and Ephrussi, "Differentiation of eye pigments," p. 230.

clear-cut cases of nonautonomous development, vermilion and cinnabar (*cn*); one ambiguous case, claret; and a dozen or so cases of genes that slightly modified the expression of others (figure 7.4).[26]

The most striking and unexpected results came with vermilion and cinnabar. To see whether these two genes were involved in the same developmental reaction, Beadle and Ephrussi arranged a reciprocal

26. Beadle and Ephrussi, "Differentiation of eye pigments," p. 230; and Beadle, "Genes and chemical reactions," pp. 1716–17.

transplant of a *cn* disk in a *v* host and a *v* disk in a *cn* host. If the products of the *v* and *cn* genes—called the v^+ and cn^+ substances—were the same, then the expression of the *v* or *cn* genes should not be altered by the host. If the v^+ and cn^+ substances were different, the transplanted eye should be wild type in both cases, since the host could make up for the chemical deficiency of the mutant eye tissue. Beadle and Ephrussi anticipated only these two logical possibilities; hence their amazement when a *cn* disk in a *v* host gave a cinnabar eye, but a *v* disk in a *cn* host gave a wild-type eye! The simplest explanation of this asymmetrical result was that the v^+ and cn^+ substances were chemical intermediates in a connected chain of chemical reactions, $ca^+ \rightarrow v^+ \rightarrow cn^+$ (the third intermediate, ca^+, was inferred from similar evidence from claret flies).[27] Unexpectedly, transplantation enabled Beadle and Ephrussi to perceive a developmental relationship that was not revealed by purely genetic analysis. *Drosophila* began to be something more than a genetic instrument.

The $ca^+ \rightarrow v^+ \rightarrow cn^+$ reaction chain was in effect the blueprint of an experimental system, analogous to neo-Mendelian formulas and gene maps. It, too, was a kind of map, but unlike chromosome maps it did not represent physical locations but functional, developmental relationships. This representation of developmental reactions was an invitation to Beadle and Ephrussi to go further in transforming *Drosophila* from a genetic instrument into a developmental and biochemical one.

The transformation was gradual, however. Beadle and Ephrussi expended an extraordinary amount of labor and ingenuity working out the effects of moderating eye-color genes on the $ca^+ \rightarrow v^+ \rightarrow cn^+$ chain— to an extent that now seems almost extravagant. They constructed a sensitive double-recessive test fly that could register the slightest moderating effects on eye color. In thousands of transplants they systematically explored the minutiae of a genetic system that grew more and more complicated the more they knew about it. By the time they had exhausted this line of work, after three years, Beadle and Ephrussi had constructed a system in which no fewer than fifteen genes were shown to participate, besides vermilion and cinnabar (figure 7.5).[28] This pic-

27. Beadle and Ephrussi, "Differentiation of eye pigments," pp. 242–44; B. Ephrussi and G. W. Beadle, "Development of eye colors in *Drosophila:* Transplantation experiments on the interaction of vermillion with other eye colors," *Genetics* 22 (1937): 65–75; and Beadle and Ephrussi, "Development of eye colors in *Drosophila:* Diffusable substances and their interrelations," ibid., 76–86, on pp. 80–81.

28. Beadle and Ephrussi, "Development of eye colors," p. 85.

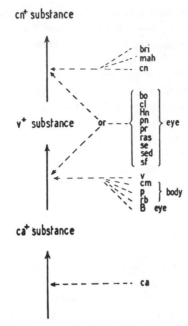

·Diagram indicating assumed relations of various genes to the three diffusible sub
stances. Dotted lines indicate the step assumed to be interfered with in the various mutant types,
in body, eye, or both. The particular alleles indicated in diagram are not necessarily the ones
used in the experiments.

Figure 7.5   Relationships between genes and intermediates in the synthesis
of eye-color pigments, as inferred from transplantation data. From G. W.
Beadle and B. Ephrussi, "Development of eye colors in *Drosophila:* Diffusable
substances and their interrelations," *Genetics* 22 (1937): 85.

ture was a geneticist's representation of a developmental system, not
an embryologist's.

Collaboration and Careers

The predominance of genetics in Beadle and Ephrussi's work can also
be seen in the social relations of their collaboration. Beadle, though
junior to Ephrussi in years, seems to have been the senior partner in
the period of their most intense collaboration in 1935–37. It was not
because he was a domineering person; quite the contrary, Beadle displayed remarkable tact and sensitivity when handling Ephrussi's de-

manding and temperamental (if charming) personality. More likely it was *Drosophila* itself that enabled Beadle to set the course of the collaboration. *Drosophila* was a genetic system, and so it was Beadle's genetic interests and know-how that set the agenda, not Ephrussi's expertise in embryology and *Entwicklungsmechanik*. Had the two decided instead to invent a genetics of sea urchins—Ephrussi's creature—a collaboration would undoubtedly have had a quite different dynamic.

Ephrussi did try to apply transplantation to morphological development, but his high hopes were always dashed. He and Beadle transplanted disks of wings, legs, and other organs, hoping to find cases of nonautonomous development. Ephrussi also planned to make interspecies hybrids using transplantation, and to revive his work on lethal mutants to study the genes that controlled early embryonic development.[29] But, alas, every organ system developed normally, quite unaffected by the host tissues. Ruth Howland and her group of *Drosophila* embryologists at New York University had the same experience, and their early enthusiasm for transplantation and genetics soon waned, as did Ephrussi's.[30] Between unproductive embryological work and productive genetic work the choice was easy.

The astonishing productivity of their eye-color work—thirty papers, many highly original, in just three years—earned Beadle and Ephrussi rapid career advancement. In 1935 Ephrussi was promoted to associate director of the tissue culture laboratory, which was devoted increasingly to the transplantation project. In 1937 he was named Maître de Recherche of the Centre Nationale de Recherche Scientifique, as well as director of a new Laboratory of Genetics at the École des Hautes Études. There Ephrussi commanded a *Drosophila* laboratory with four full-time workers, plus technicians. Bright young biologists began to be attracted to the transplantation work, most notably the young Jacques Monod. In 1936 Ephrussi struck up a collaboration on

29. Burian et al., "Singular fate," pp. 397–400; A. Mayer to W. E. Tisdale, 2 Mar. 1936, pp. 3–4, and supplement III, pp. 2–4, RF 1.1 500D 12.127.7; Demerec to Ephrussi, 13 Nov. 1936, MD; and Beadle and Ephrussi, "Transplantation in *Drosophila*."

30. R. B. Howland, B. P. Sonnenblick, and E. A. Glancy, "Transplantation of wing-thoracic primordia in *Drosophila melanogaster*," *Amer. Nat.* 71 (1937): 158–66; C. W. Robertson, "The metamorphosis of *Drosophila melanogaster*, including an accurately timed account of the principle morphological changes," *J. Morphol.* 59 (1936): 351–99; and Glancy and Howland, "Transplantation of mutant and wild type bristle-bearing tissues in *Drosophila melanogaster*," *Biol. Bull.* 75 (1938): 99–105.

the chemistry of eye pigments with Yvonne Khouvine, a bioorganic chemist. For a young scientist in France it was a meteoric ascent, almost unheard of.[31]

As for Beadle, *Drosophila* transplantation earned him an offer from Harvard in 1936 and a counteroffer from Morgan of an assistant professorship. (Beadle turned Harvard down at first, because their "enthusiasm for *Drosophila* work was a bit weak," but accepted a renewed bid.) Just a year later he became a full professor at Stanford, where he proceeded to set up a *Drosophila* transplantation lab. With a grant of $3,000 from the Rockefeller Foundation, he hired a bacterial biochemist, Edward L. Tatum, to work full-time on the chemistry of eye-pigment precursors.[32] The intense collaboration between Beadle and Ephrussi was more or less over by 1937. Transatlantic collaboration was difficult to keep up when both partners were responsible for large working groups.

Independently, though still in frequent touch by mail, Beadle and Ephrussi began to move the transplantation work from genetics to biochemistry, together with the chemists Tatum and Khouvine. They made this move when they were growing discouraged with the work on genic interactions, which neither extended the chain of reactions nor shed light on the physiology of each step. Indeed, the picture became less clear the more was known about it. Beadle and Ephrussi were embarrassed when claret turned out to be not a step in the developmental chain but merely a modifier of the vermilion–cinnabar reaction, and even worried for a time that the v^+ and cn^+ substances might after all be only different amounts of the same thing.[33]

Beadle and Tatum were also pushed in the direction of biochemistry by the addition to their experimental repertoire of the biochemical

31. Roman, "Boris Ephrussi"; Mayer to Tisdale, 2 Mar. 1936, supplement III, pp. 2–4, Ephrussi, "Report 1937–38", Fauré-Fremiet memo, 13 July 1938, and Ephrussi report, 25 Apr. 1939, p. 8, all in RF 1.1 500D 12.127; Ephrussi to Demerec, 4 Oct. 1936, MD; and Andre Lwoff, "Recollections of Boris Ephrussi," *Somatic Cell Genetics* 5 (1979): 677–79.

32. Beadle to Demerec, 15 Feb., 13 Apr. 1936, MD; Joshua Lederberg, "Edward Lawrie Tatum," *Biog. Mem. Nat. Acad. Sci.* 59 (1990): 357–86; C. V. Taylor to Weaver, 5 Apr. 1937, Hanson to R. L. Wilbur, 3 May 1937, Hanson diary, 6 May 1937, 13–23 Apr. 1938, and Weaver diary, 11 Mar. 1936, 27 Jan. 1937, 28 Jan. 1939, all in RF 1.1 205D 8.103–107; and Taylor to Weaver, 23 Aug. 1938, RF 1.1 205D 10.135.

33. Beadle and Ephrussi, "Differentiation of eye pigments"; and G. W. Beadle, "The development of eye colors in *Drosophila* as studied by transplantation," *Amer. Nat.* 71 (1937): 120–26.

techniques of transfusion and nutritional bioassay. These apparently minor changes in technique transformed *Drosophila* more profoundly than transplantation ever did. Techniques borrowed quite innocently from biochemistry led Beadle further than he ever expected, across the threshold from genetics and embryology into biochemistry.

Drosophila Transformed: Transfusion and Feeding

What were these experimental techniques that so changed the life of Beadle's lab? Transfusion was a simple variant of transplantation, in which body fluid, or lymph, instead of imaginal disks, was transferred by injection from one larva to another. In the feeding method, chemical substances were fed to larvae in their food, rather than physically injected. The purpose was the same as transfusion: to expose developing tissues in eye-color mutants to suspected intermediates in the synthesis of eye pigments. However, feeding had consequences that were farther reaching and less predictable. First, it meant importing biochemists into genetics labs: biochemists had developed the feeding method with microorganisms and rodents, and they had the know-how, if anyone did, to extend the method to *Drosophila*. But biochemists brought with them their own perceptions of how *Drosophila* should be used as a biochemical tool. They redesigned *Drosophila* for work on nutritional biochemistry and, as a result, chemical problems became more productive and competitive with genetic problems. New players, with different research agendas and craft skills, impelled drosophilists toward biochemical lines of experimental work.

Transfusion entered Beadle and Ephrussi's project inconspicuously in 1935–36 as a supplement to transplantation, to confirm that pigment precursors were indeed synthesized in specific organs and transported via the lymph to the developing eye.[34] So used, transfusion seemed to be a minor variant of transplantation. However, it reversed the relation between implant and host. In transplantation experiments the implanted disk was modified by the host; it was the object tested. Thus used, *Drosophila* was a genetic instrument, a biological host into which genes were inserted, though by physical means, now, rather than by genetic recombination. In transfusion experiments, in contrast, the

---

34. B. Ephrussi, "Chemistry of 'eye color hormones' of *Drosophila*," *Quar. Rev. Biol.* 17 (1942): 327–38, on pp. 327–28, 331–32. A *v* or *cn* fly developed a normally pigmented eye if at the pupal stage it was injected with fluid extracted from wild-type pupae.

host fly was itself the object tested, registering the effects of added chemicals. Transfused flies were used in the same way that rats or microorganisms were used by biochemists to register the effects of vitamins or essential food factors. It was a fundamental difference, though not perceived as such at first. The change from transfusion to feeding likewise seemed to be a small step. Both had the same purpose; feeding was merely simpler than injection, less traumatic to larvae, and more amenable to quantitative assay. It, too, was subversive, however, because even more than transfusion it caused *Drosophila* to be used in ways that were more typical of biochemists than of geneticists.

These new methods quietly transformed *Drosophila* into a biochemical tool. As Ephrussi suggestively put it, transfused flies were chemical "reagents" for detecting tiny amounts of biologically active substances.[35] Having such a good instrument for biochemical diagnosis was a powerful incentive for Beadle and Ephrussi to do biochemistry with it. Transfusion and feeding gave them the means and the incentive to perfect *Drosophila* as a tool of biochemical assay. Thus did apparently minor variations in experimental technique set off major changes in how *Drosophila* was perceived and used.

Ephrussi and Beadle took different approaches to their new line of work. Ephrussi's characteristic foraging strategy was opportunistic, quick and dirty. In mid 1936 he and Khouvine transfused randomly selected amino acids into vermilion larvae in the hope that one would turn out to be the v^+ substance. This shortcut to identifying active substances was a standard practice of nutritional biochemists, but it was shooting in the dark. When none of the compounds proved able to make vermilion eyes into wild-type, Khouvine and Ephrussi turned to other things.[36]

In contrast, Beadle and his colleagues worked slowly and systematically, reconstructing *Drosophila* into a new kind of standard system. Beadle began chemical work at Harvard in collaboration with the biochemist Kenneth Thimann, an old friend from Caltech, and Lloyd W. Law, who was finishing his Ph.D. in genetics.[37] At Stanford, Beadle set

35. Ephrussi, "Chemistry of 'eye color hormones,'" pp. 330–32.

36. Ephrussi to Demerec, 30 Nov. 1936, MD; Ephrussi to Beadle, 18 Oct. 1937, GWB 1.26; Ephrussi report, 25 Apr. 1939, RF 1.1 500D 12.127; and Y. Khouvine, B. Ephrussi, and M. H. Harnly, "Extraction et solubilité des substances intervenant dans la pigmentation des yeux de *Drosophila melanogaster,*" *Comptes Rendus Acad. Sci. Paris* 203 (1936): 1542–44.

37. K. Thimann and G. W. Beadle, "Development of eye colors in *Drosophila:* Extraction of the diffusible substances concerned," *Proc. Nat. Acad. Sci.* 23

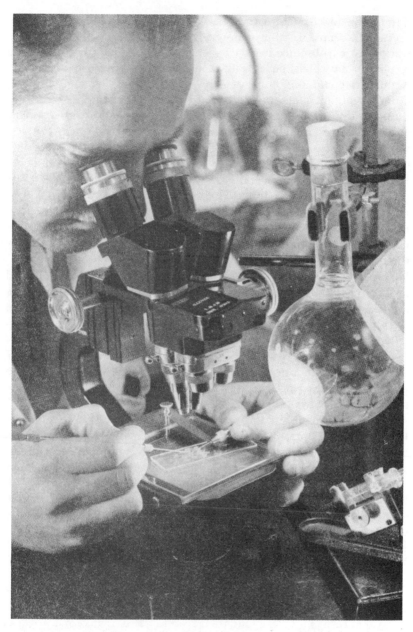

Figure 7.6   George Beadle performing a transplantation experiment on *Drosophila* larvae, circa 1940. Courtesy of American Philosophical Society Library, Stern Papers, photographs file Beadle.

Tatum the task of improving Law's assay and making the feeding method into a precision instrument, a "reagent" for measuring concentrations of v^+ substance in fly extracts. Tatum first constructed special test stocks in which, because they had virtually no eye pigment, the slightest deepening of color would be visible. Then, by compounding different mutant genes, he constructed a set of nine standard reference flies, so that extracts from test flies could be compared with standard colors and given a numerical value. Here indeed was a reagent, a new standard *Drosophila* system for efficient, quantitative, large-scale biochemical assay.[38] It was like Bridges's standard mapping stocks, but constructed for a different purpose.

Tatum's bioassay required a good deal of practical craft skill to make it work. There were many little ways to go wrong, which biochemists had the experience and know-how to avoid, but which were unseen pitfalls for the uninitiated. Ephrussi discovered this when he tried for the first time to make it work. "I should be *very, very* grateful to you," he wrote Beadle, "for giving me a short description of your adopted procedure (temperature? how do you raise them [larvae]? At what age are the different controls used for comparison? Can you keep one set of control flies for some time? etc.)—as fast as possible."[39] It took detailed instructions and some trial and error at the bench to make *Drosophila* work as well as a chemical assay in Paris as it did in Palo Alto. The feeding method was new and tricky. Even Beadle hesitated to go public with a description of it. (Ephrussi suggested that he limit the risk of exposure by publishing in a "more or less neglected journal"!)[40]

Tatum's elegant bioassay revived Ephrussi's interest in the chemistry of eye-color hormones. Beadle told him about it in August 1937, and in October Ephrussi and Khouvine began their first experiments using reagent stocks supplied by Tatum.[41] Their first project, after they learned to make the system work, was to repeat their experiments with amino acids. Some remarkable effects turned up almost at once. Most

(1937): 143–46; Beadle and L. W. Law, "Influence on eye color of feeding diffusible substances to *Drosophila melanogaster*," *Proc. Soc. Exp. Biol. Med.* 37 (1938): 621–23; and Ephrussi to Beadle, 4 Aug. 1937, GWB 1.26.

38. Tatum and Beadle, "Development of eye colors in *Drosophila*: Some properties of the hormones concerned," *J. Gen. Physiol.* 22 (1938): 239–53.

39. Ephrussi to Beadle, 2 June 1938; see also 5 Dec. 1938, 4 Mar. 1939; all in GWB 1.26.

40. Ephrussi to Beadle, 29 Nov., 30 Dec. 1937, GWB 1.26.

41. Ephrussi to Beadle, 7 Aug., 18, 22 Oct. 1937, GWB 1.26.

of the test larvae died of starvation on a diet of sugar and a few amino acids. However, two of those that did hatch had eyes of a distinctly deeper color, indicating that the amino acids they had been fed were chemically related to the v^+ hormone. Ecstatic, Ephrussi repeated the experiment with still larger numbers of larvae, and got more positive results—but with a different set of amino acids! Nothing was repeatable; there was no relation between specific amino acids and pigment formation. Ecstasy turned to puzzlement, and puzzlement to despair. Only a few weeks after thinking he had cracked the chemistry of v^+ at the first try, Ephrussi was wondering if the whole business was not "a big mistake." It was "a hell of a mess," he wrote Beadle, "difficulties came from the most unexpected sides and mixed up everything." The effect was finally revealed to be an artifact of the experimental method.[42] Embryological experiments proved just as disappointing. For a time Ephrussi was in a state of high excitement, thinking he had found a hormone that controlled the growth of eye facets; it too, however, proved to be an experimental phantom.[43] Years of hard labor thus came to nothing—a stark contrast with the effortless productivity of the transplantation work.

Beadle and Ephrussi talked of resuming their collaboration to work on the chemistry of eye-color hormones, but they never did. They were no longer mobile postdocs, and they both had chemical co-workers who were accustomed to the winner-take-all style of nutritional biochemistry and were not enthusiastic about collaboration. So Beadle and Ephrussi agreed to work independently, swapping technical know-how and sending each other copies of manuscripts before publication to avoid treading on each other's toes. Beadle supplied Ephrussi with

42. Ephrussi to Beadle, 18, 22 Oct., 8 Nov., 26 Dec. 1937, GWB 1.26. Ephrussi finally had to agree with Beadle and Tatum that his new phenomenon was a phantom, an artifact caused by starvation, when starved tissues diverted to the synthesis of eye pigments amino acids that would normally have gone to make proteins. Ephrussi, "Chemistry of 'eye color hormones,'" pp. 332–34; Y. Khouvine, B. Ephrussi, and S. Chevais, "Development of eye colors in *Drosophila:* Nature of the diffusible substances; effects of yeast, peptones, and starvation on their production," *Biol. Bull.* 75 (1938): 425–46; Tatum and Beadle, "Effect of diet on eye-color development in *Drosophila melanogaster,*" *Biol. Bull.* 77 (1939): 415–22; and Ephrussi to Beadle, 21, 25 Mar., 8 June 1938, GWB 1.26.

43. Ephrussi to Beadle, 19 July, 4 Aug., 26 Dec. 1937, 14 Feb. 1938, 17 Apr., 21 July 1939, 25 July 1941, GWB 1.26; and B. Ephrussi, Y. Khouvine, and S. Chevais, "Genetic control of a morphogenetic substance in *Drosophila melanogaster?" Nature* 141 (1938): 204–5.

Tatum's latest bioassay technology, while Ephrussi provided Beadle with quantities of *Calliphora* pupae, which were the best source of crude hormone extract. Differences in experimental results and interpretations were discussed in almost weekly letters. It was, as Ephrussi noted, "a form of cooperation, and perhaps the most convenient under the present conditions."[44]

Ephrussi somewhat ungenerously blamed his troubles on the absence of real collaboration with Beadle.[45] In fact, his difficulties seem mainly due to his excessive eagerness for dramatic results and his incautious use of an unfamiliar and quite tricky experimental system. Ephrussi seemed never certain when an experiment was complete and safe to publish. *Drosophila,* used as a biochemical system, had hidden pitfalls for people in a hurry.

Beadle and Tatum's more deliberate and calculated style of work also had its pitfalls, as became apparent when Tatum was scooped in the search—by then a race—to identify the chemical structure of the v^+ substance. Their chief competitors were Alfred Kühn and his group, who worked on the eye-pigment system on the meal moth, *Ephestia kühniela.* Tatum seemed clearly in the lead in 1939; he had a pure, crystalline derivative of the v^+ substance and was working day and night to identify the active principle.[46] Then, through two bizarre but serendipitous mistakes in experimental technique, Ephrussi and Tatum revealed that the v^+ hormone was a relative of the amino acid tryptophane. Before they could capitalize on their lucky break, however, the prize was snatched from their grasp by Kühn and the eminent organic chemist Adolf Butenandt, who was experienced in races to solve chemical puzzles. Knowing from Tatum and Ephrussi's work what to look for, Butenandt had only to test every chemical relative of tryptophane until the right one turned up.[47]

44. Ephrussi to Beadle, 9 Jan. 1937 [*sic:* 1938]; see also 8 May, 26 Dec. 1937, 25 Mar., 8 May, 2, 13, 16, 29 June, 4 Nov., 5 Dec. 1938, 4 Mar., 23 Oct. 1939; all in GWB 1.26.

45. Ephrussi to Beadle, 21 July 1939, GWB 1.26. He also blamed Khouvine; by 1939 he had decided to replace her with a more experienced organic chemist, when the outbreak of war put an end to all his plans: Ephrussi to Breadle, 15 Oct., 5 Dec. 1938, 4 Mar., 17 Apr., 21 July 1939, GWB 1.26.

46. H. Miller diary, 1 June 1939, RF 1.1 500D 12.127.

47. Ephrussi, "Chemistry of 'eye color hormones,'" pp. 327–38; E. L. Tatum and G. W. Beadle, "Effect of diet on eye-color development in *Drosophila melanogaster,*" *Biol. Bull.* 77 (1939): 415–22; Tatum, "Development of eye colors in *Drosophila:* Bacterial synthesis of v⁺ hormone," *Proc. Nat. Acad. Sci.* 25 (1939):

Being scooped by the Germans was a bitter pill, especially for Tatum, who had worked so hard to perfect *Drosophila* as a biochemical reagent. But that is how life was in nutritional biochemistry, a notoriously volatile and risky game in which years of labor could be wiped out in an instant by an interloper's lucky guess. By reconstructing *Drosophila* as a biochemical instrument, Beadle and Ephrussi moved into an area of scientific practice where different and not entirely congenial customs prevailed. In the transplantation work Beadle and Ephrussi had virtually no competitors; possession of a new experimental system enabled them to skim the cream off the problem. In nutritional biochemistry, however, promising leads were easily exploited by others, and rewards went to the lucky or the quick-footed opportunist. For those who staked their careers on being the first to discover new vitamins or hormones, life was a zero-sum game: only one person got a prize, the one who got it first. It was a jarring contrast with drosophilists' customs of reciprocity, disclosure, and avoidance of head-on competition, and Beadle was repelled by the "bitch the other guy if you can" attitude of the biochemists.[48]

It was in this context that Beadle conceived the idea of introducing a new creature into the laboratory, *Neurospora*.

The Invention of Biochemical Genetics

The muse of experiments descended upon Beadle unexpectedly while he was sitting one day in early 1941 listening (or not listening) to Tatum lecturing on comparative biochemistry.[49] It was a simple idea that stole into his mind: Rather than starting with visible mutations and laboriously working out their biochemistry, why not start with biochemical mutations and work out their genetics? It would mean, however, abandoning *Drosophila* for a microorganism, a creature in which mutations that blocked the different steps in a biosynthetic chain of reac-

---

486–90; Ephrussi to Beadle, 17 Apr. 1939, GWB 1.26; and A. Butenandt, W. Weidel, and E. Becker, "Kynurenine als Augenpigmentbildung auslösendes Agens bei Insekte," *Naturwissenschaften* 28 (1940): 63.

48. Beadle to F. J. Ryan, 4 Jan. 1943, GWB 2.23; see also Beadle to E. Brand, 4 Jan. 1942, GWB 1.7.

49. Notes taken by Carleton Schwerdt indicate that Tatum lectured on the nutrition of fungi on 18 Feb. 1941, though that is not necessarily when Beadle had his brainstorm: Joshua Lederberg, personal communication.

tions would be visibly expressed as new nutritional requirements (for the products of the blocked reactions).[50] Tinkering with *Drosophila* would not do, because there was no practical way of screening for metabolic mutants with *Drosophila*. A new creature would have to be constructed to perform a new variety of biochemical genetics.

But exactly what place did Beadle anticipate the new organism would have in his lab? It has generally been assumed that Beadle meant *Neurospora* to displace *Drosophila;* that he was fed up with the eye-pigment project and was looking for a new line of work to replace it. Certainly, Beadle left that impression, calling the Germans' coup "a blessing in disguise."[51] Yet a closer look at contemporary perceptions gives a different picture. To most observers Butenandt's stunning success with the v^+ substance seemed to open doors, not close them. Tatum got right to work on the cn^+ substance, and Ephrussi, having escaped by the skin of his teeth from the German occupation of France, resumed work on the pigment problem as soon as he was safely relocated at Johns Hopkins in 1941. Nothing, he thought, stood in the way of a complete chemical elucidation of the chain of reactions between tryptophane and eye pigments, and he looked forward to reviving "the old cooperation" with Beadle. Sturtevant was no less optimistic than Beadle appeared about the prospects of the work.[52] It does not seem that the *Drosophila* work had reached a point of "erratic and diminishing returns," as one historian has put it.[53] So why *Neurospora?*

The puzzle is deepened by the fact that Beadle clearly did not expect to give up *Drosophila* himself. He had just begun a big new project on genetic dominance in the eye-color series. By combining classical

50. G. W. Beadle, "Biochemical genetics: Some recollections," in John Cairns, Gunther S. Stent, and James D. Watson, eds., *Phage and the Origins of Molecular Biology* (Cold Spring Harbor, N.Y.: Cold Spring Harbor Laboratory, 1966), pp. 23–32, on p. 29; Beadle, "Recollections," pp. 7–8; and Beadle, "Genes and chemical reactions," p. 1717.

51. Beadle, "Genes and chemical reactions," p. 1717; Lederberg, "Tatum," p. 2; and G. W. Beadle, "Chemical genetics," in L. C. Dunn, ed., *Genetics in the Twentieth Century* (New York: Macmillan, 1951), 221–39, on pp. 224–25.

52. Miller diary, 1 June 1939, p. 3, RF 1.1 500D 12.127; Ephrussi, "Chemistry of 'eye color hormones,'" pp. 336–37; Ephrussi to Beadle, 23 Oct., 4 Dec. 1939, 19 Feb., 2 May 1940, 18 Apr., 25 July 1941, GWB 1.26; Sturtevant, "Physiological aspects of genetics," *Ann. Rev. Physiol.* 34 (1941): 41–56, on p. 48; and G. W. Beadle and E. L. Tatum, "Experimental control of development and differentiation," *Amer. Nat.* 75 (1941): 107–16.

53. Kay, "Selling pure science," p. 80.

genetic analysis with Tatum's new bioassay, Beadle hoped to understand dominance in quantitative, physiological terms.[54] He seemed in
no way discouraged by the potential of *Drosophila* for studying the
chemistry of gene action. So, why *Neurospora?*

The answer, I think, is that Beadle meant *Neurospora* to be more
Tatum's system than his own. That is, he introduced *Neurospora* not to
replace the eye-pigment work but to advance a new line of work that
Tatum was inaugurating on the nutritional biochemistry of insects. Evidently it did not get very far before being supplanted by the far more
successful work on *Neurospora:* it resulted in no publications and was
never mentioned by either Tatum or Beadle. Indeed, the very existence
of this line of experiment can only be inferred from circumstantial
evidence. Nevertheless, it seems crucial to understanding how *Drosophila* was unexpectedly dislodged from its place of honor in Beadle's laboratory.

What is the evidence that Tatum was planning to make *Drosophila*
into a standard organism for doing nutritional biochemistry in insects?
First, he did far more work on the nutrition of *Drosophila* than seems
strictly necessary for his bioassay of eye pigments. Second, he explicitly
pointed to insect nutrition as an important and underworked field,
likening it to animal nutrition on the eve of the discovery of vitamins. It
was the unmistakable gesture of a man marking a piece of intellectual
territory for his future use. Other biochemists were making similar
signals: for example, the eminent Harvard biochemist Yellapragada
SubbaRow.[55] It would have been a logical step for Tatum, since he knew
the nutritional physiology of *Drosophila* as well as its genetics, which
almost no other biochemist did. To make *Drosophila* into a standard
nutritional organism he had only to discover, by a process of elimination, what chemicals larvae needed to grow on a chemically defined,
minimal medium. It was a task he had performed many times before
with bacteria and fungi.[56]

But would Beadle, an old hand in turning organisms to new tasks,
have shared Tatum's optimism for insect nutrition? What were the
likely returns? Eye color was really the only character of *Drosophila* that

54. Beadle to Ephrussi, 17 Nov. 1941, GWB 1.26.

55. E. L. Tatum, "Nutritional requirements of *Drosophila melanogaster,*" *Proc.
Nat. Acad. Sci.* 25 (1939): 490–97; and Tatum, "Vitamin B requirements of *Drosophila melanogaster,*" ibid., 27 (1941), 193–97.

56. E. L. Tatum, H. G. Wood, and W. H. Peterson, "Growth factors for bacteria, V: Vitamin B, a growth stimulant for propionic acid bacteria," *Biochem. J.*
30 (1936): 1898–1904.

Figure 7.7   Edward L. Tatum inspecting *Neurospora* cultures (1960s?). Courtesy of Rockefeller Archive Center, Edward L. Tatum Papers.

was suited to biochemical analysis, and it would have been nearly impossible to make nutritional mutants of *Drosophila*, since they would only be recognized by their failure to live on minimal media. Practically speaking, metabolic mutations would be lethals in *Drosophila*, which, unlike microorganisms, died when starved. It is understandable why

the idea popped into Beadle's mind that it would be less risky to construct the genetics of a microorganism than to find metabolic mutants
in *Drosophila*.

Beadle had personal as well as scientific reasons for interceding in
Tatum's program of research. Tatum's career at Stanford was at a critical point in late 1940. He was still a research associate on soft money,
despite three years of devoted and highly productive work. Beadle had
been trying to get Tatum a faculty appointment, but some of the Stanford biologists balked at diverting their resources to a chemist. Beadle
could hardly do without Tatum, but he could not in good conscience
keep him much longer in a dependent position.[57] (Just like Morgan
and his boys!) Beadle felt morally responsible for Tatum, having en-
ticed him away from a safe career in nutritional biochemistry into the
no-man's-land between genetics and chemistry, where it was all too easy
to fall between academic stools. Beadle was forceably reminded of this
danger when Tatum's father, a professor of pharmacology, took him
aside and expressed his concern about Edward's prospects "in a position in which he is neither a pure biochemist nor a bona fide geneticist."[58]

In fact, Tatum was made an instructor in biology at Stanford a few
months later, but the point to note is the conjunction of these events
in Tatum's career with Beadle's invention of biochemical genetics. Was
it a coincidence? Maybe so, but given Beadle's skill as a strategist of
careers, it is safer to assume that he knew what he was doing when he
conceived a mode of practice that would make use of Tatum's micro-
biological skills and avoid the risks of redesigning *Drosophila* to an end
for which it seemed ill-suited. Hence, *Neurospora*.

Neurospora: Constructing an Instrument

What was this creature, *Neurospora*, that displaced *Drosophila* from
Beadle's laboratory? In the wild *Neurospora* is an unruly interloper. It
grows irrepressibly in tropical countries like Indonesia, where botanists
first studied its natural history. There, because its spores are heat resistant, it was the first colonizer of terrains devastated by fire or volcanic
eruptions. The tough, fast-growing creature covered landscapes of

57. Beadle to Weaver, 18 Jan. 1940; Hanson to R. J. Williams, 8 Oct. 1940;
Hanson diary, 30 Dec. 1940; all in RF 1.1 205D 10.135.

58. Beadle, "Recollections," pp. 7–8; and Weaver diary, 17 Feb. 1941, RF
1.1 205D 10.135.

blackened stumps and plant debris with a blanket of bright orange fuzz. *Neurospora* also lived in a symbiotic relationship with humankind as a participant in an important agricultural technology. It was widely used by the Javanese to brew *ontjom,* a favorite concoction made by covering thin cakes of peanut mash (the residue of oil pressing) with leaves and allowing it to collect a thick orange covering of the fungus. *Neurospora* resisted the more rigorously domesticated regimen of the scientific laboratory, however. The Dutch botanist Friedrich Went, who first investigated the fungus in Java in the late 1890s, was disconcerted by its ability to grow right through the cotton plugs of his culture tubes. He could do nothing with it until he got it back to the more temperate North. Yet even there European strains of *Neurospora* were trouble-makers. Bakers dreaded infestations of the red bread mold *N. sitophila,* whose heat-resistant spores passed unharmed through ovens inside loaves and thus survived to invade and ruin the next day's work.

The natural history of *Neurospora* was thus not unlike *Drosophila*'s. *Neurospora,* too, was a hanger on of mankind, flourishing in their houses and backyards on the abundance of agricultural products and by-products of pressing, preserving, and fermenting. It was a more deliberate participant in agricultural technology than *Drosophila,* which participated in fermentation only inadvertently, by carrying yeasts from one place to another on its little feet. Nor did *Drosophila* trouble anyone as *Neurospora* troubled bakers, but both creatures inhabited the domesticated, "second" nature of humankind, from whence it was but a short step into experimental laboratories.

Neurospora had in fact been a very junior partner to *Drosophila* in Morgan's group for more than a decade before Beadle picked it up. It had been tamed and domesticated to life in the lab in the 1920s, mainly by Bernard O. Dodge, a plant pathologist at the Brooklyn Botanical Garden. Dodge revealed that the creature came in two mating types and had a primitive sex life, and therefore a genetics. Dodge was a Columbia Ph.D. and a regular visitor in Morgan's group, and his discovery made him into a flag-waving promoter of *Neurospora* as an organism for genetic research—even better than *Drosophila,* he claimed. Morgan was not persuaded, but he did agree to take a collection of Dodge's cultures with him to Caltech, and there a few years later Carl C. Lindegren adopted it for his dissertation subject and worked out its cytogenetics. Dodge followed the fly group and *Neurospora* west in 1934, when he took a position at the University of Southern California, in Los Angeles. *Neurospora* was thus no stranger to the drosophilists, though far from a favorite. Ever the opportunist, it was at the right place at the right time when Beadle began to cast about for a microorganism that had both a biochemistry and a genetics.[59]

Neurospora required some redesigning and rebuilding to make it into an instrument of biochemical genetics. Dodge and Lindegren had already worked out standard procedures for measuring linkage, and Lindegren was able to tell Beadle all the little tricks of *Neurospora* culture, such as how to get spores to germinate. *Neurospora*'s nutritional requirements were not known, and Tatum had to work them out. They proved to be quite simple: besides a source of carbon and nitrogen and salts, just biotin, a B vitamin. As an opportunistic survivor in the wild, *Neurospora* could not have afforded to be a fastidious eater. Its propensity to grow also had to be controlled, since detecting small amounts of metabolic intermediates required that growth rates be measured quantitatively. Initially, colonies were simply scraped off their agar plates, squeezed dry, and weighed, but that was clumsy, and Francis J. Ryan devised a better method, which made use of four-foot glass "race tubes" in which the progress of the leading edge of the mycelium could be marked at different times and a growth rate calculated. (Ryan, who came to Stanford as a National Research Council fellow to work on embryology, was so taken with *Neurospora* that he loitered around the lab until Beadle finally agreed to take him in.)[60]

Simple procedures for producing and identifying mutants were also worked out. Filaments were irradiated and then allowed to fuse with untreated filaments of the opposite sex. From each of the resulting fruiting bodies (perithecia) a single spore pod was dissected out, split open, and its eight spores gingerly removed and placed in order on an agar plate—four potential mutants and four wild types. Each spore was then transferred to a growth tube with enriched growth

59. Beadle, "Genetics and metabolism in *Neurospora,*" *Physiol. Rev.* 25 (1945): 643–63, on pp. 644–45; C. L. Shear and B. O. Dodge, "Life histories and heterothallism of the red bread-mold fungi of the *Monila sitophila* group," *J. Agric. Res.* 34 (1927): 1019–42; and Beadle, "Genes and chemical reactions," pp. 1717–18.

60. C. C. Lindegren to Beadle, 17 Mar., 15 Dec. 1941, 20 Jul 1942, GWB 1.41; Beadle to Lindegren, 3 Feb. 1943, GWB 1.41; Beadle, "Genetics and metabolism in *Neurospora,*" *Physiol. Rev.* 25 (1945): 643–63, on pp. 644–50; Arnold W. Ravin, "Francis J. Ryan (1916–1963)," *Genetics* 84 (1976): 1–25, on pp. 4–5; F. J. Ryan, G. W. Beadle, and E. L. Tatum, "The tube method of measuring the growth rate of neurospora," *Am. J. Bot.* 30 (1943): 784–99; and Tatum and Beadle, "The relation of genetics to growth-factors and hormones," *Growth* 6 suppl. (1942): 27–37.

medium, and from each tube a bit of epithelium was transferred again to a tube with minimal medium. Mutants that were blocked in one or another biosynthetic pathway would not grow and so could be identified by simple inspection. To find out which biosynthetic pathway was blocked and at what intermediate step, it was then simply a matter of adding back to the minimal culture chemicals that were known or suspected intermediates and seeing which ones restored growth. In this way chains of biochemical reactions could be constructed.[61]

Beadle and Tatum knew the method would work, since it was all tried-and-true nutritional practice; the only question was, Would biochemical mutations appear so infrequently that the mutant hunters would quit in despair before finding one? So Beadle and Tatum agreed to isolate 1,000 cultures before testing them for mutants. It was an unnecessary precaution. The first mutant (for pyridoxine) turned up in the 299th culture and the second (for thiamine) in 1,090. Just six months after their first experiment Beadle wrote Ephrussi that "the mutants are turning up so fast we haven't got time to digest them yet." By November 1941 he and Tatum had a dozen "good" mutants. Within another year they had tested 33,000 single-spore cultures and identified 83 mutants. By 1945 over 60,000 tests had turned up mutants in 100 different genes having to do with metabolic processes. Beadle's team of mutant hunters, now a dozen strong, produced more knowledge of biosynthetic pathways than two or three generations of biochemists had with their traditional methods.[62]

The productivity of the *Neurospora* method astonished everyone. It was like the early *Drosophila* work all over again, and the flood of new mutants that poured out of the mutant-hunting room at Stanford had the same ability to inspire awe and wonder. One colleague professed not to believe that Beadle had done all he claimed—how could he have done so much so fast?[63] In fact, Beadle's mutant-hunting room

61. Beadle, "Genetics and metabolism," pp. 646–50; G. W. Beadle and E. L. Tatum, "*Neurospora*, II: Methods of producing and detecting mutations concerned with nutritional requirements," *Am. J. Bot.* 32 (1945): 678–86; and A. M. Srb and N. H. Horowitz, "The ornithine cycle in *Neurospora* and its genetic control," *J. Biol. Chem.* 154 (1944): 129–39.

62. Beadle to Ephrussi, 1 Aug., 17 Nov. 1941, GWB 1.26; Beadle to Lindegren, 25 July 1941, GWB 1.41; Beadle, "Genes and chemical reactions," p. 1717–18; Beadle, "Progress report," 24 Sept. 1942, RF 1.1 205D 10.142; Beadle, "Genetics and metabolism in *Neurospora*," p. 652; and N. H. Horowitz, "Fifty years ago: The *Neurospora* revolution," *Genetics* 127 (1991): 631–35.

63. Ephrussi to Beadle, 5 Jan. 1942, GWB 1.26; and Beadle, "Biochemical genetics: Some recollections," p. 29.

was organized for mass production, just as Morgan's fly room had been. The process of screening mutants lent itself to assembly-line production, and results poured forth faster than anyone could have anticipated. Norman Horowitz, Beadle's chief mutant hunter, recalled how every day brought exciting new results—it was a "scientific paradise."[64] Back home in New York, Francis Ryan felt like an exile from that paradise: "On hearing of the 12,000–16,000 culture tubes weekly," he wrote, "we in New York, the proverbial hub of the universe, feel provincial."[65] Here was an experimental practice so novel and so productive that it put every competitor in the shade. Including *Drosophila*.

The extraordinary productivity of *Neurospora* completely upset the ecology of Beadle's group—it was as vigorous a colonizer of new terrain in the laboratory as it was in the wild. Within months, work on the *Drosophila* eye-color system was put on the back burner, then abandoned. Beadle never meant this to happen. In August 1941 he seemed to share Ephrussi's hope that *Neurospora* would not lead him to give up *Drosophila*.[66] But by November he reluctantly had to admit that even his new line of work on the physiology of genetic dominance would have to be set aside. It was a bitter pill. He had just finished constructing various combinations of eye-color mutants and was eager to test them with Tatum's bioassay. But he realized he would probably never get back to *Drosophila*: "The Neurospora work is constantly accelerating, and I find I'm more and more tied to it," he wrote Ephrussi. "This state of affairs seems to apply to all the 'fly lab' workers, so Drosophila seems to be at least temporarily out of luck at Stanford."[67] The nascent collaboration with Ephrussi on eye pigments was another casualty of the *Neurospora* invasion. Beadle gently suggested that Ephrussi might strike up a collaboration with someone else, but Ephrussi was politely unenthusiastic.[68] Beadle and Tatum and their group of biochemists and geneticists never went back to *Drosophila*. A few years later Ephrussi as well dropped *Drosophila* to work on the genetics of yeast.

Neurospora entered Beadle's lab in much the same way that *Drosophila* had entered Morgan's. *Neurospora*, too, came in by the side door,

64. Beadle to Hanson, 24 Feb. 1942 (with "Report"), 3 Apr. 1943, RF 1.1 205D 10.142–143; Beadle to Ephrussi, 1 Aug. 1942, GWB 1.26; and Horowitz, "Beadle," pp. 35–36.

65. Ryan to Beadle, n.d. [1942?], 27 June 1942, GWB 2.23.

66. Ephrussi to Beadle, 22 Aug. 1941, GWB 1.26.

67. Beadle to Ephrussi, 17 Nov. 1941, GWB 1.26.

68. Beadle to Ephrussi, 1 Aug. 1942; and Ephrussi to Beadle, 25 July 1941, 4 Aug. 1942; GWB 1.26.

not to displace dominant organisms and their lines of work but, if my reconstruction is right, to inaugurate an additional line of experiment for which it seemed uniquely well suited. However, *Neurospora* did unexpectedly displace other organisms from their central privileged places in the domestic economy of the lab. And it did so in precisely the way that *Drosophila* had displaced *Phylloxera* and mice in Morgan's lab in 1910, by revealing a capacity for producing lots of very useful mutants, not useful for mapping chromosomes, now, but for mapping biochemical pathways.

The different roles of Beadle and Tatum in the invention of *Neurospora* offers a no less striking parallel with the roles of Morgan and his "boys" in the invention of *Drosophila*. As *Drosophila* mapping was designed not for Morgan himself but as a project for his young students, so, too, was mapping biochemical pathways mainly a project for Tatum and his team of young biochemists. As Morgan was eventually pushed to the margins of his own laboratory, so, too, was Beadle, as Tatum's biochemical mapping project expanded and took over. As genetic mapping was the creation of young people and a game for the young, so, too, was biochemical genetics and molecular biology: simple and fast-paced, narrower biologically than the mode of practice it replaced, but far more productive. Thus did *Drosophila* and the genetics of mapping and transmission lose their place of honor in Beadle's group to *Neurospora* and the construction of metabolic pathways. Beadle cared far less for the *Neurospora* work than Tatum and his group of biochemists, though he recognized its value and allowed it to dominate his group. Within a few years Beadle had more or less abandoned active research and become an administrator and Morgan's successor as head of the Biology Division of Caltech.[69]

Developmental Genetics Revisited

Another casualty of *Neurospora*'s invasion was Beadle and Ephrussi's original interest in development. In retrospect they had left embryology behind even earlier, when they first undertook the search for the v^+ substance and began to fashion *Drosophila* into a reagent for biochemical assay. One of the few people to see this was Richard Goldschmidt, who pointed it out amidst the chorus of praise for the trans-

---

69. For Beadle's later career see Lily E. Kay, *The Molecular Vision of Life: Caltech, the Rockefeller Foundation, and the Rise of the New Biology* (New York: Oxford University Press, 1993), chap. 7.

plantation work. "The Beadle-Ephrussi work ... is certainly most outstanding and admirable work," he wrote Schultz, "but I wonder whether the final outcome has not made it a complete flap in regard to Physiological Genetics. There can be no doubt that the [stuffs?] are chromogen precursors and nothing else and decidedly no hormones. Thus the interest lies mainly in the chemistry of the pigment formation but genetically it's just a parallel to what we know rather thoroughly from plant pigments. Am I wrong?"[70] Goldschmidt did not grasp the novelty and growth potential of what Beadle and Ephrussi had done, but he was certainly right that it had little interest for people like himself, who dreamed of connecting genetics and development. A new mode of experimental practice had been invented, even a new discipline, which some people were already calling "molecular biology" and which had absolutely nothing to do with development.

The separation of genetics and development was fundamentally a consequence of Beadle and Tatum's choice of *Neurospora* and the way they constructed and used it. Unbeatable for charting metabolic pathways, the fungus was useless for studying development, as Beadle discovered when he tried to use it in that way. For example, he avidly collected mutants with different colony shapes in preparation for a study of the biology of growth and cell morphology, but was embarrassed when his "mutants" turned out to be contaminating fungi that had exploited his ignorance of mycology to find a home in his *Neurospora* stockroom.[71] Beadle also tried to develop the cytology and physiology of *Neurospora,* but it was slow work and could not compete with the exciting, fast-paced biochemical mapping.[72]

It was not just that *Neurospora,* a simple microorganism, had no real embryology. Every aspect of the system into which it was built helped turn it to biochemical uses: the fact that Tatum, a biochemist, was in charge of the mutant-hunting work; the existence of a large constituency of biochemists who were eager consumers of new knowledge of biochemical pathways; practical and financial pressures from war-research agencies, who wanted quick new biochemical assays for vitamins—everything seemed to impel Beadle to use *Neurospora* for bio-

70. R. Goldschmidt to Schultz, 17 Dec. 1942, JS.

71. Tatum and Beadle, "Genetic control of biochemical reactions," pp. 30, 34; Beadle to Hanson, 17 Jan. 1944, RF 1.1 205D 10.144; and Beadle, "Recollections," pp. 8–9.

72. Beadle memorandum, 18 Dec. 1941, and Beadle to Hanson, 14 Feb. 1942, 2 Dec. 1944, all in RF 1.1 205D 10.141–142, 144; and B. McClintock to Beadle, 9 Jan., 12, 22 Aug., 27 Nov. 1944, 8 May 1945, GWB 2.5.

chemical purposes.[73] In the 1910s similar imperatives of the production process drove drosophilists to develop their creature's potential for genetic mapping at the expense of its developmental and evolutionary aspects. And so it was again with *Neurospora* and biochemical mapping.

Neurospora may have driven *Drosophila* out of Beadle's workshop, but in laboratories where *Drosophila* reigned without competitors, developmental geneticists continued their search for new experimental systems. In the late 1930s several lines of work were begun that mark the real beginning of embryological genetics in *Drosophila*. Unlike the developmental work of the 1920s, these new kinds of experiment dealt with embryological processes and with the genes that determined morphogenesis. These new experimental methods were true hybrids of genetics and embryology, and unlike earlier improvisations they actually did produce knowledge of how embryogenesis worked.

It is not surprising to find such a widespread burst of activity in developmental genetics, since Sturtevant, Schultz, Demerec, and others had been promoting it for over a decade. The question is, Why did their efforts suddenly begin to bear fruit around 1936 or so? One reason is that the basic embryology of *Drosophila* was finally worked out and published. It was accomplished in a remarkably short time, mostly between 1930 and 1935, by a handful of people: at Caltech by T. Y. Chen and Dobzhansky; at Leningrad by J. J. Kerkis, and at the University of Frankfurt by Marie Strasburger, H. Hertweck, H. Rühle, and R. Gleichauf. Ruth Howland's group of insect embryologists at New York University also contributed.[74] This accumulation of basic embryological knowledge made it possible to use *Drosophila* as an instrument for experimental embryology. No longer was it necessary to improvise with trick systems that had been designed for chromosomal mechanics. Drosophilists could finally begin to do what advocates of developmental genetics had long dreamed of.

At Caltech, for example, Donald F. Poulson began a systematic investigation into the effects of various chromosomal deficiencies on the embryogenesis of *Drosophila*. The idea was to find out where the genes

73. Kay, "Selling pure science"; Robert E. Kohler, "Systems of production: *Drosophila, Neurospora,* and biochemical genetics," *Hist. Stud. Phys. Sci.* 22 (1991), 87–130, on pp. 123–27; and Beadle to Hanson, 15 Apr., 13 July 1942, RF 1.1 205D 10.142.

74. For a useful summary and bibliography see D. F. Poulson, "The effects of certain X-chromosome deficiencies on the embryonic development of *Drosophila melanogaster,*" *J. Exp. Zool.* 83 (1940): 271–321.

were located that controlled morphogenesis. (Ideally one would remove one gene at a time, but blocks of genes were more practicable to deal with.) Poulson also planned studies of the effects of chromosomal deficiencies on the rate of metabolism in developing eggs and on the embryogenesis of nerve fibers, to localize the points of action of developmental genes.[75]

It also became possible for the first time to study mutants of the genes that regulated early differentiation, mutants like *aristapedia* and *proboscopedia,* which caused legs to grow where antennae or mouth parts normally did, or *bithorax* of later fame, which disrupted the process of segmentation and caused an extra thoracic segment to be formed. Until the basic embryology of *Drosophila* had been done there was little that could be done with these monsters, but in the late 1930s they became practical tools for studying the genetics of development.[76] Beadle and Ephrussi's transplantation technique also opened up new opportunities for experimental manipulation of developing embryos. Although cases of nonautonomous development of disks remained disappointingly rare, transplantation did prove fruitful in studies of histogenesis. For example, Ernst Hadorn and Curt Stern used transplantation in their study of the development of the reproductive apparatus, out of which came the notable discovery of a small endocrine organ (the ring gland) that controls the onset of pupation.[77]

Obviously, such experiments were nothing like Beadle and Tatum's biochemical genetics. They were biologically more complex and far less productive systems, evolving slowly and painfully as know-how gradually accumulated. Such experiments were more like traditional experimental morphology than the fast-paced genetics, to which drosophilists had grown accustomed. So slow and undramatic was the

75. D. F. Poulson, "Chromosomal control of embryogenesis in *Drosophila,*" *Amer. Nat.* 79 (1945): 340–63; Poulson, "Chromosomal deficiencies and the embryonic development of *Drosophila melanogaster,*" *Proc. Nat. Acad. Sci.* 23 (1937): 133–37; Poulson to Demerec, 11 Feb. 1938, 26 June, 30 Oct. 1939, 14 May 1940, MD; Sturtevant to Demerec, 24 Apr. 1936, MD; and Poulson to Metz, 24 May 1936, MD file Streeter.

76. Claude Villée, "Phenogenetic studies of the homoeotic mutants of *Drosophila melanogaster,*" *Amer. Nat.* 79 (1945): 246–58.

77. Stern to Beadle, 29 July 1936, Beadle to Stern, 23 Aug 1936, Stern to Dobzhansky, 13 Sept. 1937, Dobzhansky to Stern, 26 Sept. 1937, all in CS; Stern, reports to F. B. Hanson, n.d. [1937], 26 Jan., 2 June 1938, n.d. [1941], CS file RF; and E. Hadorn, "An accelerating effect of normal 'ring-glands' on puparium-formation in lethal larvae of *Drosophila melanogaster,*" *Proc. Nat. Acad. Sci.* 23 (1937): 478–84.

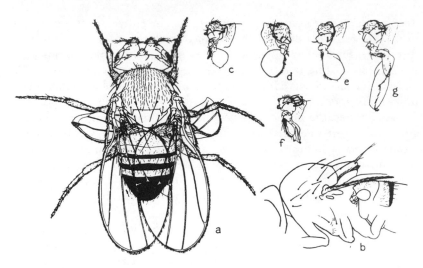

Figure 7.8 *Bithorax* mutant, used to study the genetic control of development. Inset *b* shows the wild type, and insets *c–f* show variations in the bithorax character, with extra wings in the place of balancers and an additional thorax-like structure. From Morgan, Bridges, and Sturtevant, "Genetics of Drosophila," fig. 37, p. 79.

development of these systems, in fact, that *Drosophila* became thoroughly unfashionable in the 1950s and 1960s. It seemed hopelessly old-fashioned in comparison with *E. coli* and the bacteriophages, and drosophilists found research grants hard to come by. Painstaking work on development could not compete with the brilliant novelties and amazing productivity of molecular genetics. Lines of work in developmental genetics, begun in the late 1930s and early 1940s by Donald Poulson, Edward B. Lewis, Ernst Hadorn, and others, only became really fruitful in the 1970s, when new techniques were devised for studying the master genes that regulate differentiation. Then *Drosophila* came once again into the scientific limelight.[78]

78. E. B. Lewis, "Genes and gene complexes," in Alexander Brink, ed. *The Heritage of Mendel* (Madison: University of Wisconsin Press, 1967), pp. 17–47; M. P. Scott, "Segmentation and homoeotic gene function in the developing nervous system of *Drosophila,*" *Trends in Neuroscience* 11 (1988): 101–6.

Conclusion

The emergence of a new mode of developmental genetics in the late 1930s reveals how laboratory sciences are shaped by the practical imperatives of experimental work. For first-generation drosophilists, the sheer abundance of standard *Drosophila* made developmental genetics a poor competitor with chromosomal mechanics. Boundary crossing was also inhibited by the need to keep genetic and developmental phenomena analytically distinct, lest geneticists and embryologists lose their credibility with their special audiences. Thus the process of production and publication offered few incentives for drosophilists to tinker with established boundaries and experimental practices. For developmental genetics to be revived, drosophilists had to fashion problems and practices that were both genetic and developmental, appealing to constituencies on both sides of the boundary. They had to venture outside their customary work culture and construct experimental systems that could generate as many doable problems as did older, standard systems.

Beadle was above all else a highly original inventor of novel transdisciplinary experimental systems: transplantation, transfusion, feeding, *Neurospora*. Like his mentors in the fly group, Beadle liked fast-paced, productive modes of experimental work. The rapid succession of experimental practices in his laboratory—from classical to developmental genetics and then to biochemical genetics—was in large part a result of his willingness to venture outside the confines of mainstream *Drosophila* practices. Trespassing enabled him to escape the limitations that had been built into drosophilists' experimental systems in the early years of genetic mapping.

Beadle's willingness to cross disciplinary boundaries was encouraged by a widespread programmatic commitment to reuniting genetics with development and evolution. Where this new ideology came from and why it gathered strength around 1930 remain unanswered questions, but its strong local manifestation at Caltech obviously reflected Sturtevant and Morgan's broad biological range, the fly group's habit of collective work, and their determination to be in the vanguard of whatever was new in "Drosophilistics." A generational dynamic may also have encouraged foraging outside the mainstream. Around 1930 a second generation of drosophilists was entering a more crowded discipline, and for these newcomers it made sense to seek career advantages at the margins, where genetics rubbed elbows with other modes of experimental biology. It is no accident, I think, that the movement to put the gene back in the whole organism coincided with the appearance of new centers of practice and new communication infrastructure like *DIS*. Interdisciplinarity was also encouraged by financial and administrative trends in academia, notably the increase in research funding by private foundations, whose officers, to avoid being trapped by disciplinary lobbies, chose to support transdisciplinary projects.[79]

The succession of Beadle's experimental systems also reveals how the process of change can be driven by seemingly small and innocuous changes in experimental practices. Transplantation was meant to (and did) make *Drosophila* into an instrument of developmental genetics, but it also set Beadle on a course that eventually took him beyond development to molecular genetics. A desire to improve transplantation led to the introduction of transfusion and feeding methods that were seemingly modest improvements on transplantation but ended up overturning the ends to which *Drosophila* was used. *Neurospora* was introduced, if I am right, to improve Tatum's chances of success in genetic biochemistry, but it proved so unexpectedly productive that it overshadowed genetics in Beadle's laboratory and drove *Drosophila* out of it altogether. Why do apparently innocent experimental techniques have this transforming power? In part because experimental scientists operate in a social and moral system that rewards production, and because they follow problems opportunistically. Also, experimental organisms and practices carry more baggage than is sometimes visible: the built-in experiences and past choices of entire disciplines. So when Beadle and Tatum adopted the standard practices of nutritional biochemists, and an organism that was suited to these practices, they also willy-nilly subscribed to their agendas.

Transdisciplinary collaborations, in which hybrid practices are born, are also likely to be subversive of established disciplinary customs. Morgan worried that Schultz had lost interest in genetics as a result of his work with Selig Hecht.[80] Beadle found himself unexpectedly converted to *Drosophila* from maize by his work with his fly-group colleagues. Ephrussi came to Caltech with the idea of adding a genetic dimension to his developmental work, but he ended up doing *la vrai génétique*. Beadle's partnership with Ephrussi led to his giving up chromosomal mechanics for developmental problems. He hired Tatum for his skills in biochemical assay, but Tatum turned Beadle's genetic program into a biochemical one.

79. Robert E. Kohler, *Partners in Science: Foundations and Natural Scientists, 1900–1945* (Chicago: University of Chicago Press, 1991), chaps. 11–12.

80. Schultz to Morgan, 3 Jan. 1929, JS.

Partnerships with new organisms also transform experimental programs and practices, as did Morgan's partnership with the fruit fly. Similarly, the introduction of *Neurospora* changed the ecology of laboratories where *Drosophila* had long been the dominant species. *Neurospora*, and later *E. coli* and phage, gradually took over the core areas of genetic mapping and gene action, for which fly, maize, and mouse had been the preferred organisms for over thirty years. *Drosophila* became an organism of choice for biologically more complex lines of work in development and evolution, but was largely excluded from the center ring of molecular genetics in the 1950s and 1960s. *Drosophila* and drosophilists thus found their domestic home range narrowed by competition at the same time that they found more diverse niches and ways of life in other domains of experimental biology. Extinction seems to be a rare event in the natural history of experimental creatures, though some creatures may come close to it.[81] The more usual fate of laboratory creatures in a world of changing scientific fashions is to be perpetually reinvented, along with their human symbionts, and refashioned for novel modes of experimental work.

81. *Neurospora*, for example, suffered a dramatic decline in use and status when *E. coli* became the standard creature of molecular biologists. See also Gregg Mitman and Anne Fausto-Sterling, "Whatever happened to *Planaria*? C. M. Child and the physiology of inheritance," in Adele E. Clarke and Joan H. Fujimura, *The Right Tools for the Job: At Work in Twentieth-Century Life Sciences* (Princeton: Princeton University Press, 1992), pp. 172–97.

EIGHT

From Laboratory to Field: Evolutionary Genetics

BOUT THE TIME OF BEADLE AND EPHRUSSI'S BREAKTHROUGH
with transplantation, other drosophilists were inventing practices that reunited genetics with experimental evolution. Here,
too, major changes occurred when experimenters learned to get
around the practical limitations that drosophilists had built into their
experimental system in the early years of genetic mapping. Evolution
is essentially about variation and variability in populations; thus the essential feature of experimental evolution is a practical method of dealing with variability in whole populations. But as we saw in previous
chapters, drosophilists' standard systems worked so well precisely because they confined or eliminated the natural variability of wild drosophilas. Mapping and chromosomal mechanics required standard
flies, and having laboriously constructed standard tools, drosophilists
had every reason to get the benefits from their investment. There was
little incentive to return to the messy variability of undomesticated
flies. Yet that was precisely what had to happen for genetics and experimental evolution to be reunited. Drosophilists had to leave their laboratories and go afield. They had to see variability not as a nuisance
but as a significant and doable problem. They had to invent standard lab
and field methods for dealing with large numbers of nonstandard,
wild flies.

The most successful of these new experimental systems was invented by Theodosius Dobzhansky between 1936 and about 1942.
What transplantation was to developmental genetics, Dobzhansky's genetics of natural populations was to evolutionary genetics. With the
methods he invented, drosophilists could do experiments on the mechanics of variation and speciation in natural populations. Dobzhansky
succeeded where Sturtevant, Metz, and others had failed twenty years
before, but he did so in a quite different and unexpected way. Dobzhansky's mode of practice was no improvisation but a new experimental system, one so productive and expandable that it became the basis
of a whole new discipline in the 1940s and 1950s.

One reason why Dobzhansky's mode of practice was so fruitful is that it combined field practice with the theoretical biometrics of J. B. S. Haldane, R. A. Fisher, and Sewall Wright. In the 1940s and 1950s, for example, Dobzhansky engaged in an extraordinarily intense and fruitful collaboration with Sewall Wright, Wright providing the design principles for Dobzhansky's field and laboratory experiments. This work has been analyzed and celebrated as a centerpiece of the so-called evolutionary synthesis.[1] I will not deal with the theoretical side of Dobzhansky's work here, in part because it has been expertly analyzed by others, but also because I believe the turning point in evolutionary genetics occurred somewhat earlier, when drosophilists left the laboratory for the field and invented practical methods for doing experiments with large and diverse populations of wild flies. At that stage theory was, I think, not important. Rather, practical changes in experimental work created opportunities for new injections of theory.

I will focus here on the crucial boundary between laboratory and field, indoors and outdoors. Fieldwork was not incidental to Dobzhansky's new experimental practice, as it was for most drosophilists, but constitutive. He and his disciples made regular field trips to forests, mountains, and deserts. They integrated field collecting with laboratory practice in equal measure. Whereas classical geneticists mapped chromosomes and followed the movements of genes in crossing-over, Dobzhansky produced geographical-genetic maps of regional and local populations of drosophilas and charted changes in the genetic constitution of natural populations that were produced by changes in climate, competition, and other selective pressures. This new mode of genetic mapping connected genetic makeup not to the dynamics of chromosomes and transmission in individuals but to the dynamics of speciation in natural populations.

Although Dobzhansky's mode of field and lab practice was the prevalent one in evolutionary genetics in the 1940s, Dobzhansky was neither the only nor the first drosophilist to cross the boundary between lab and field. In the mid 1920s Sergei Chetverikov and his group of able students in Moscow analyzed the genetic load of mutations in local populations of several local species of wild drosophilas, as well as

1. William B. Provine, "Origins of *The Genetics of Natural Population* series," in Richard C. Lewontin, John A. Moore, William B. Provine, and Bruce Wallace, eds., *Dobzhansky's Genetics of Natural Populations, I–XLIII* (New York: Columbia University Press, 1981), pp. 5–79; Lewontin, "The scientific work of Th. Dobzhansky," ibid., pp. 93–115; Ernst Mayr and William B. Provine, *The Evolutionary Synthesis* (Cambridge: Harvard University Press, 1980); and Provine, "The role of mathematical population geneticists in the evolutionary synthesis of the 1930s and 1940s," *Stud. Hist. Biol.* 2 (1978): 167–92.

of a wild population of *D. melanogaster* from the Black Sea coast (far removed from any escapees from domesticated stocks). This work was continued in Berlin in the late 1920s by Nikolai and Helene Timofeeff-Ressovsky, using Berlin wild stocks of *melanogaster*. In the early 1930s in Moscow, Chetverikov's disciple D. D. Romashov continued to study genetic diversity in wild populations of various *Drosophila* species, together with N. P. Dubinin and his students.[2] In the mid 1930s John T. Patterson and his group at the University of Texas began systematic, large-scale collecting of the extremely diverse *Drosophila* species of the Southwestern United States and Central America. By the late 1930s work with novel species collected from the field was the fashion in many corners of the *Drosophila* world. Most varieties of field genetics were different from Dobzhansky's, however, and less directly concerned with evolution. The Timofeeff-Ressovskys became more interested in mutations than in evolution, and Patterson's group was mainly concerned with systematics and biogeography: that is, patterns of distribution of different species and their mechanisms of dispersal. Dubinin was more drawn to questions of adaptation and ecology than to systematics and speciation. Dobzhansky's evolutionary ideas and field practices were in fact closest to those of Chetverikov and his original group of students, perhaps because Dobzhansky, like Chetverikov, had been trained as a field entomologist and came to *Drosophila* genetics only after extensive experience as a field naturalist.[3] Dobzhansky, like his Russian predecessors, applied systematic fieldwork to study the dynamics of local populations and the mechanism of speciation. Of all the variant forms of field genetics, Dobzhansky's was the closest to the ideal of an experimental science of the evolutionary process, and that characteristic makes it an ideal case study for this book. We want to understand how drosophilists managed to reunite genetics and evolution productively after two decades of frustrating improvisations, and how

2. Mark B. Adams, "Sergi Chetverikov, the Kol'tsov Institute, and the evolutionary synthesis," in Mayr and Provine, *Evolutionary Synthesis,* pp. 242–78; Adams, "The founding of population genetics: Contributions of the Chetverikov school, 1924–1934," *J. Hist. Biol.* 1 (1968): 23–39; Adams, "Toward a synthesis: population concepts in Russian evolutionary thought, 1925–1935," *J. Hist. Biol.* 3 (1970): 107–29; Adams, personal communications; and Diane B. Paul and Costas B. Krimbas, "Nikolai V. Timoféeff-Ressovsky," *Sci. Amer.* (Feb. 1992): 86–92.

3. Adams, "Sergei Chetverikov"; and Nikolai Krementsov, "Dobzhansky and Russian entomology: The origin of his ideas on species and speciation," in Mark Adams, ed., *The Evolution of Theodosius Dobzhansky* (Princeton: Princeton University Press, 1994).

one novel hybrid mode of field and laboratory practice was shaped by the local experimental culture of the Caltech group.

A second point to be made is the intimate connection between experimental practice and the moral economy of working groups. The invasion of the Caltech fly lab by thousands of wild drosophilas dramatically changed the character of its communal life. *D. melanogaster* lost its place of eminence and was replaced by newcomers, most notably *D. pseudoobscura.* (After Bridges died in 1938, no one in the fly group worked on *melanogaster.*) Established lines of work on chromosomal mechanics were overshadowed by Dobzhansky's novel and highly productive mode of work. People who had invested in traditional systems no longer enjoyed the same unchallenged status in the group. Fundamental changes in experimental production are bound to create conflicts in a group's social relations, and the fly group was not immune to such disruption. Whereas Beadle made an early and graceful transition to an independent career, Dobzhansky suffered a falling out with Sturtevant, his closest friend and collaborator, that left lasting scars on both. It is a skeleton in the drosophilists' closet which to this day they do not like to have brought out. Sturtevant and Dobzhansky's falling out was not just a personal feud, however, but something more interesting and important. It reveals the fly group's traditional moral economy stressed and failing under the pressure of fundamental changes in its key experimental practices.

An Apprentice Drosophilist

We have seen in a previous chapter what the fly group looked like to Dobzhansky when he and his wife, Natasha, arrived there from Russia in December 1927. But what did Dobzhansky look like to the fly group? A bit lost at first, like any immigrant, unable to follow what anyone said except Edith Wallace, who spoke a clear Yankee dialect; also, idealistic, energetic, eager, and very starstruck. More than anything, Dobzhansky was determined to become a member of the *Drosophila* elite. He had already had a little experience with fly breeding. In 1922, while a teacher in the agricultural faculty of the Polytechnic University of Kiev, he had carried out a little project on the pleiotropic effects of eye-color genes, using the mutant stocks that H. J. Muller had just brought to Moscow as a gift to his new Russian friends. Beyond that, however, Dobzhansky's knowledge of *Drosophila* genetics was limited to what he could get from the fly group's outdated 1915 textbook. What Dobzhansky knew best was experimental systematics and morphology. His favorite creature was the ladybug *Coccinella*, which he collected avidly in the field, and he excelled in applying microanatomical dissection to the study of insect systematics. One of his first *Drosophila* projects in Russia was a study of the morphology of the sexual apparatus in different mutant strains. It was his skill and experience in microdissection, as much as his naive enthusiasm for *Drosophila* genetics, that led to his assimilation into the experimental culture of the fly group.[4]

Sturtevant seems to have taken an immediate and unusually strong liking to the young Russian. "He is a first-class worker," he told Stern in July 1928, "a Hell of a nice fellow, and has a good head."[5] When the Dobzhanskys had troubles with their visas, Sturtevant intervened to keep them from being deported; probably he saved their lives, since it was by then illegal for Soviet citizens to emigrate (several returning Russians were in fact executed in the purges).[6] Dobzhansky was ambitious and hard working, and Sturtevant made sure that he had problems to work on that made use of his special skills. Sturtevant seized upon his skill in microanatomy to extend the group's line of research on intersexes, and Dobzhansky soon found himself involved in a highly productive collaboration with Jack Schultz on the genetics and development of intersexes (nine papers in four years, two with Schultz).[7]

Dobzhansky also struck out on his own. Just six months after his arrival he undertook to see if Muller's new X-ray method could be used to produce translocations—at the time a rare and valuable kind of mutation. It was a characteristic move: he was impressed by the bandwagon that Muller had got rolling and wanted to be on it. His first experiments were done at Woods Hole in July 1928, and within a very short time he had produced no fewer than nine translocations. It was a first, almost—Muller published first, to Dobzhansky's chagrin—but Dobzhansky's translocations were nonetheless a gold mine for the fly group: a score of papers came out of them in the next few years. It was not long after his discovery that Morgan walked into his room and in-

4. Provine, "Origins," pp. 6–16; Theodosius Dobzhansky, oral history, Butler Library, Columbia University, pp. 238–51; and Dobzhansky, Über den Bau des Geschlechtsapparats einiger Mutanten von *Drosophila melanogaster* Meig," *Z. induk. Abstamm.* 34 (1924): 245–48.

5. A. H. Sturtevant to C. Stern, 1 July 1928, CS.

6. Dobzhansky, oral history, pp. 300–326.

7. Dobzhansky, oral history, pp. 244–47, 328; Provine, "Origins," pp. 18–20; and Th. Dobzhansky and C. B. Bridges, "Reproductive systems of triploid intersexes in *Drosophila melanogaster*," *Amer. Nat.* 62 (1928): 425–34.

Figure 8.1   Clarence Oliver and an early X-ray apparatus used for inducing mutations, University of Texas, 1927. Courtesy of Lilly Library, Indiana University, Muller Papers.

vited him to stay on at Caltech as an assistant professor.[8] He thus realized his dream of joining the company of genetic heroes (for so they seemed to him), whom he had admired for so long from afar.

By 1932 Dobzhansky had made a distinctive place for himself in the group's productive and moral economy. He rivaled Jack Schultz as a talker and intellectual gadfly.[9] He was also the group's most energetic benchworker, turning out papers twice as fast as Bridges and Sturtevant and four times as fast as Schultz. He could not bear to leave data unpublished. To him "a month gone by without a paper sent to press was a wasted month," James Bonner recalled.[10] By fly group standards Dobzhansky's urge to publish seemed an "excessive itch," but Dobzhansky was just more hard-driving than they were, more eager to get ahead. As he put it to Demerec, Bridges's point of view was that "there is no rush (to which I reply that the only thing that is becoming better with age is wine)."[11]

In other ways, however, Dobzhansky found it hard to get used to the plain and somewhat parochial style (by European standards) of American academic culture. His American hosts, he felt, lacked the high idealism of his European mentors, and he often felt lonely and far from his real home. For Sturtevant he felt a particular admiration and friendship, yet he deplored what he saw as his lack of cultural and spiritual qualities. Lots of little things bothered him: Morgan's and Sturtevant's militant secularism, Sturtevant's Anglophilia and his dislike of mixing science and general philosophizing—so unlike the cultural style of European academics, who were spiritual exemplars and *Kulturträger*, bearers of high culture.[12] Americans pursued science more as a middle-class occupation than as a sacred calling. Dobzhansky liked their lack of pretense but always seemed most comfortable socially with fellow Europeans like Demerec, Karl Belar, Peo Koller, or Curt Stern.

This cultural difference may explain Dobzhansky's curious silence on the subject of species and evolution. From his experience as a field

8. Dobzhansky, oral history, pp. 282–83, 294–99; Th. Dobzhansky, "Translocations involving the third and the fourth chromosomes of *Drosophila melanogaster*," *Genetics* 15 (1930): 347–99; H. J. Muller to E. Altenburg, 14 July 1928, HJM-A; and Muller to Stern, 3 Oct. 1929, CS.

9. Dobzhansky to M. Demerec, 30 Sept., 18 Oct., 1 Nov. 1933, 7, 16 Jan. 1934; Demerec to Dobzhansky, 8 Nov. 1933; all in MD.

10. Provine, "Origins," pp. 48–49.

11. Ibid; and Dobzhansky to Demerec, 16, 21 Feb. 1934, MD.

12. Dobzhansky to Stern, 9 Apr. 1933, CS; and Dobzhansky, oral history, pp. 251–55, 269–70, 439.

entomologist and from his Russian mentors he had acquired an abiding interest in evolutionary mechanisms and a distinctive view of species not as a homogeneous type but as a congeries of different local populations. He wrote of these things in Russian and discussed the evolutionary import of his translocations in letters to his teacher, Iurii Filipchenko.[13] But in English it was just the facts: no philosophizing.

Exactly what role his Russian connections played in the invention of Dobzhansky's variety of evolutionary genetics remains an open question, which research now in progress in Russian sources may soon answer.[14] In any case, it seems clear that his knowledge of the work of Chetverikov and his group was not a crucial element in the sequence of local events that led Dobzhansky in the early 1930s from classical chromosomal mechanics back to evolution and fieldwork. He was familiar with Chetverikov's ideas and results but seems not to have consciously taken Chetverikov's fields methods, or his ideas, as a model for his own work.[15]

Dobzhansky did not just sit down and consciously decide to invent a new experimental system, as Beadle and Ephrussi did. Rather, it seems that short-term opportunities for productive work led him from one thing to another, step by step, imperceptibly from classical genetics to a new mode of experimental field genetics. This gradual process may have been partly a result of Dobzhansky's unusually strong desire to produce unceasingly and to produce work that would win praise from his American mentors: that meant not departing too abruptly from the mainstream. But there is also a more general point: experimentalists do not usually give up a productive system of work until they have another, equally productive system in hand. Programmatic beliefs do encourage foraging and enable experimenters to recognize new things when they stumble upon them. But experimenters change course when doable practices make this possible with minimal sacrifice

13. Adams, "Sergei Chetverikov," p. 268; Krementsov, "Dobzhansky and Russian entomology; Provine, "Origins," pp. 8–13; and Daniel Alexandrov, personal communication.

14. Mark B. Adams and Daniel Alexandrov, personal communication.

15. Dobzhansky later said that he had not jumped on "the Chetverikov bandwagon" earlier because he was busy with translocations and other things, and besides it seemed a "very small bandwagon" at the time: Dobzhansky, oral history, pp. 335–36, 409–11. He also followed the work of Romaschoff and Dubinin as it was published (in Russian) in the early 1930s and knew what the Moscow drosophilists were doing from Russian visitors to Caltech: Mark B. Adams, personal communication.

Figure 8.2   Theodosius Dobzhansky, Natasha Dobzhansky, N. I. Vavilov, and
G. D. Karpechenko at Caltech, 1932. Courtesy of American Philosophical
Society Library, Dobzhansky Papers.

of production. New practices evolve gradually as elements of established practices are modified and rearranged, in a way that is highly
contingent on the particularities of local working groups. Dobzhansky
made his way from *D. melanogaster* and chromosomal mechanics to wild
D. pseudoobscura and evolution by following his practical, experimentalist's nose.

From Hybrid Sterility to Obscurology

The key connecting link in the succession of Dobzhansky's work
was hybrid sterility, as Will Provine has shown.[16] Hybrids led him to *D.
pseudoobscura*, and *pseduoobscura* led him out of the laboratory and into
the field. Dobzhansky became interested in the problem of hybrid sterility while working with Jack Schultz on intersexes. His starting point
was a bright idea about the mechanism of nondisjunction, which pro-

16. Provine, "Origins," pp. 21–30.

duces intersexes: Dobzhansky thought it might be caused by transloca-tions blocking the physical attraction between homologous chromo-somes and disrupting orderly pairing and separation. It then occurred to him that this simple mechanical model might explain why hybrids of *melanogaster* and *simulans* were sterile. Schultz liked Dobzhansky's idea, and together they devised a clever experimental test using a trip-loid hybrid, which, since it contained two chromosomes with identical translocations, should pair normally and result in a fertile hybrid. Schultz and Dobzhansky succeeded in constructing the triploid hybrid, but its offspring, alas, were sterile. Hybrid sterility, like intersexuality, seemed to be produced by a complex interaction of many hidden genes.[17] Dobzhansky and Schultz therefore planned a straightforward genetic analysis of the hypothetical "sterility genes." Like previous hy-brid hunters, they were drawn back to mainstream genetic practice when clever shortcuts failed.

This change in experimental tactics required a different system, however, since the *melanogaster-simulans* hybrids were useless for genetic analysis, their offspring being sterile. Schultz and Dobzhansky turned instead to Donald Lancefield's hybrid of races A and B of *D. pseudo-obscura*. Lancefield had been interested in hybrid sterility but was not at the time doing much about it. Despite a quick intelligence he lacked *Sitzfleisch*, and in the years that followed the migration of the core fly group to Caltech he gradually lost interest in research.[18] *Pseudoobscura* was thus available and ready to hand; Lancefield had partially mapped it, and Schultz had done some work with it himself a few years earlier, in connection with a study of X-ray mutation.[19]

Pseudoobscura soon began to reveal some unexpected and remark-able qualities. Strains collected from different locales turned out to have four different types of Y chromosome, three of which seemed to have been derived from the fourth by deletions and rearrangements. Hybrids constructed from different stocks of A and B also displayed a

17. Dobzhansky, oral history, pp. 329–33; Th. Dobzhansky, "The decrease of crossing-over observed in translocations and its probable explanation," *Amer. Nat.* 65 (1931): 214–32; Dobzhansky and T. Schultz, "Triploid hybrids between *Drosophila melanogaster* and *Drosophila simulans*," *J. Exp. Zool.* 65 (1933): 73–82; and Dobzhansky and Schultz, "The distribution of sex-factors in the X-chromosome of *Drosophila melanogaster*," *J. Genet.* 28 (1934): 349–86.

18. Dobzhansky to Demerec, 18 Jan. 1933, 17 Dec. 1932, MD; Provine to D. Lancefield, 26 Dec. 1979 (with Lancefield's annotations), WP; and Provine, notes of conversation with R. Boche, 20 Dec. 1979. WP.

19. Schultz to Stern, 12 Feb., 2 Aug. 1930, CS.

marked variability in the size of their testes, which could be precisely measured and used as a quantitative indicator of degree of sterility. (Testes were a convenient character for this purpose because in *pseudoobscura,* unlike *melanogaster,* their bright orange color was clearly visible through a transparent sheath.) There seemed to be no end to the tricks that *pseudoobscura* could do: one stock had odd sex ratios, another displayed what seemed to be maternal inheritance, and so on. Dobzhansky exclaimed to Demerec in December 1932 that he was "having more fun than I ever had." A month later he was doing *pseudoobscura* full time, and by January 1934 he was no longer keeping even a single stock of *melanogaster,* so completely had it been replaced in his affections by his new friend.[20]

Pseudoobscura was so full of new tricks because it was an undomesticated, wild organism that had not yet been stripped of its natural genetic diversity, as *melanogaster* had long since been. But then why had Lancefield and Sturtevant not observed all that years before? The answer, it appears, is that they had worked with just a few strains, which had been partially cleaned up and standardized for mapping and comparison with *melanogaster.* Dobzhansky, in contrast, worked with many different strains collected in the field. But why did he not simply get standard stocks on the *Drosophila* exchange? Why did he take the trouble to go into the field for new ones? Because, it seems, only stocks of race A were available (hybrids being by then out of fashion). To get stocks of race B, Dobzhansky had no choice but to go into the field and collect free-living flies. It was thus rather by accident that wild pseudoobscuras came indoors with all their natural genetic diversity intact.

Since it was known that race B lived in the Pacific Northwest (not in Pasadena, inconveniently), Dobzhansky delegated Robert D. Boche, one of his students who lived in Seattle, to trap some wild flies during his summer vacation. In the fall of 1932 Boche (he pronounced it "Bowie") brought back seven examples collected from different locales, and it was in these stocks that Dobzhansky found the genetic diversity that so astonished him. The next summer Dobzhansky himself gathered many more wild stocks. "I went for a short trip to Oregon,"

20. Dobzhansky to Demerec, 17 Dec. 1932, 18 Jan., 6 June, 22 Aug., 30 Sept., 1 Nov. 1933, 7 Jan. 1934, MD; Dobzhansky to Stern, 29 Aug., 1 Nov. 1933, CS; Dobzhansky, oral history, pp. 331–32; Sturtevant to F. B. Hanson, 21 Jan. 1935, RF 1.1 205D 7.87; Th. Dobzhansky and R. D. Boche, "Intersterile races of *Drosophila pseudoobscura* Frol.," *Biol. Zentralb.* 53 (1933): 314–30; and Dobzhansky "The Y-chromosomes of *Drosophila pseudoobscura,*" *Genetics* 20 (1935): 366–76.

he reported to Stern, "and brought [back] '57 varieties' of Drosophila pseudoobscura."[21] It was reflexive for field entomologists to gather numerous specimens to represent the full range of variability, and habits formed in Dobzhansky's early training as a collector of ladybugs evidently came to the fore. It was not one creature or a few that entered Dobzhansky's laboratory but an extended family, and that was crucial.

It was almost three years, however, before Dobzhansky recognized the potential of this natural diversity for studying population dynamics. Meanwhile, he spent most of his energy trying to identify and locate "sterility" genes, or solving the intriguing genetic puzzles that turned up, like maternal inheritance and sex ratio. In short, he and Schultz worked with *pseudoobscura* as they would have with *melanogaster*. Indeed, they had to recapitulate with *pseudoobscura* the fly group's earlier domestication and reconstruction of the standard fly. Before any mapping of sterility genes could be done, wild pseudoobscuras had to be cleaned up and standardized. Their basic cytology had to be worked out, for example. Dobzhansky found that interesting at first, but within a few months he was tired of it and looking forward to getting into the field again to collect more wild flies. He also had to build an apparatus for producing mutants by X-raying, and to do the hard work of constructing standard mapping stocks, with proper markers and cleaned-up chromosomes. That he did not look forward to, and he cut as many corners as he could. As he later confessed, his mapping was "distinctly skimpy."[22] Working out the physiology and reproductive behavior of *pseudoobscura* was another essential preliminary to mapping sterility genes, which gave Dobzhansky more headaches.[23] Dobzhansky just did not love classical genetic analysis in a visceral, tactile way, as Bridges and Sturtevant did. He seemed always glad of a chance to get out of the lab and into the field.

The mapping project seemed to drag on and on. By December 1934 Dobzhansky thought he had all the material in hand for a final assault on the problem of hybrid sterility, but by April 1935, when he was halfway through the necessary crosses, he was not so sure. Like the sex-determining genes he and Schultz had sought in vain in their work on intersexes, the genes causing hybrid sterility also eluded precise ge-

21. Dobzhansky to Stern, 29 Aug. 1933, CS; Provine, "Origins," pp. 28–30; and Dobzhansky, oral history, pp. 329–32.

22. Dobzhansky, oral history, 361–62; Dobzhansky to Demerec, 7, 16 Jan., 21 Feb., 2 Apr. 1934, MD; and Dobzhansky to Stern, 1 Nov., 7 Oct. 1933, CS.

23. Dobzhansky to Demerec, 21 Feb. 1934, MD; and Dobzhansky to Hanson, 8 Feb. 1935, RF 205D 1.1 7.87.

netic analysis. A year later Dobzhansky could only conclude tentatively that such genes did exist on the X chromosome and perhaps on chromosome 2: a meager reward for so much labor. He had produced a mountain of data, most of which would never be published, and he hated wasting effort.[24] Dobzhansky's experience recapitulated Charles Metz's, when Metz failed to find hybrids and was forced to do evolutionary genetics the hard way, by mapping. Dobzhansky was no fonder of such work than Metz had been. Thus far, work in the field had been confined to annual collecting expeditions; the main work with *pseudoobscura* was in the traditional laboratory mode. So long as Dobzhansky's main objective was hybrid sterility, he had no cause to go farther afield or to change traditional practices.

Shifting Partnerships

The unexpected success of *D. pseudoobscura* somewhat upset the fly group's human ecology. Dobzhansky later recalled the "dead skepticism" of his colleagues when he gave up *melanogaster* for *pseudoobscura*: "Should not one stick to D. melanogaster, with its abundant chromosome markers, chromosome maps, and trick chromosomes? Are these 'races' [A and B] so different from 'races' which one can build in melanogaster, e.g., by irradiation?"[25] But who exactly was he thinking of? Morgan heartily approved of his work, and Sturtevant had joined him in it.[26] Probably Dobzhansky was thinking of Schultz. His collaboration with Schultz was in fact an early casualty of his taking up with *pseudoobscura*. There is no sign of any cooling of their friendship; Schultz just turned to other things.[27] He, of course, had a greater investment in *melanogaster* than Dobzhansky did, and never displayed any interest in going afield.

At the same time, *pseudoobscura* infused new life into Dobzhansky's collaboration with Sturtevant. It began almost immediately after Stur-

24. Dobzhansky to Demerec, 10 Dec. 1934, 19 Jan., 21 Apr. 1935, MD; Th. Dobzhansky, "Studies on hybrid sterility, II: Localization of sterility factors in *Drosophila pseudoobscura* hybrids," *Genetics* 21 (1936): 113–35; and Dobzhansky, oral history, pp. 355–59.

25. Dobzhansky to E. Mayr, 15 Dec. 1970, TD.

26. F. B. Hanson diary, 14 Nov. 1934, 21 Aug. 1936, RF 1.1 205D 7.86, 7.88.

27. Dobzhansky to Demerec, 21 Feb., 2 Apr. 1934, 19 Jan. 1935; and Demerec to Dobzhansky, 9, 26 Jan. 1935; all in MD.

tevant returned from a year in England in September 1933. Two months later they were "having a grand time working on pseudo-obscura," especially the stock with the odd sex ratio. Sturtevant took charge of the genetic analysis while Dobzhansky did the cytology— their usual arrangement. A year later Sturtevant gave Dobzhansky a crucial clue to the identity of a remarkable new species, *D. miranda*, which turned up among the flies collected that summer around Puget Sound. With its odd set of chromosomes and ability to produce hybrids with *pseudoobscura* (sterile, unfortunately), *miranda* soon became Dobzhansky's darling. He proposed at first to call it *D. sturtevantiana*—testimony to the warmth of their collaboration at that time.[28]

This renewed collaboration between Sturtevant and Dobzhansky was unusually close but otherwise typically in the fly group style. They divided the labor according to their complementary skills, Sturtevant's in genetics and Dobzhansky's in cytology, and chose problems that gave them some elbow room. Sturtevant did not intrude in Dobzhansky's main line of work on hybrid sterility but picked up on opportunities in Dobzhansky's rich material that were relevant to his particular interests. He entered into collaboration where his special skills could be deployed (e.g., in systematics), or where his prior work gave him a legitimate claim—after all, he did have a substantial investment in *pseudoobscura*. Although *pseudoobscura* had been Sturtevant and Lancefield's special creature, it was the custom in the fly group not to protect turf if someone came along with a good idea for an experiment. Certainly, Sturtevant displayed no evidence of being upset when he returned after a year to find Dobzhansky working full blast on *pseudoobscura*.[29] Likewise, Dobzhansky made no effort to keep his collection of wild *pseudoobscura* to himself. Like his translocations, his wild flies were too rich a resource for one person, and he shared willingly, as was the custom. In short, Dobzhansky and Sturtevant's collaboration observed the unspoken rules of the fly group's work culture.

In the winter of 1935–36 this status quo was disrupted when Dobzhansky and Sturtevant unexpectedly discovered that *pseudoobscura*, in addition to everything else it could do, could also be used to study

28. Dobzhansky to Demerec, 1 Nov., 10 Dec. 1933, MD; Dobzhansky to Stern, 7 Oct., 1 Nov. 1933, CS; A. H. Sturtevant and Th. Dobzhansky, "Geographical distribution and cytology of 'sex-ratio' in *Drosophila pseudoobscura* and related species," *Genetics* 21 (1936): 473–90; and Dobzhansky, "*Drosophila miranda*, a new species," *Genetics* 20 (1935): 377–91.

29. At least Dobzhansky reported no signs of distress: Dobzhansky to Stern, 7 Oct. 1933, CS.

the evolutionary history of regional populations. Here was something much bigger and potentially more contentious than hybrid sterility, maternal effects, or an odd sex ratio. This new development reawakened Sturtevant's long-cherished hopes for an evolutionary genetics, and at the same time brought into the realm of possibility the evolutionary program that Dobzhansky had imbibed from Filipchenko and Chetverikov. With stakes so high and ownership so ambiguous, conflict would have been hard to avoid.

Toward an Evolutionary Genetics

How, then, did Dobzhansky and Sturtevant get from hybrid sterility to a genetics of natural populations? It happened when they realized that stocks of *pseudoobscura* from different regions had distinctive and characteristic combinations of chromosomal inversions. These regional populations were not separated by geographical barriers, yet seemed to be in the process of evolving into distinct species, as races A and B already had to some extent.[30] This fundamental change occurred in late 1935, when Dobzhansky was with Sturtevant working up material collected that summer. In September he still seemed preoccupied with mapping sterility genes.[31] As the work progressed, however, he and Sturtevant began to notice distinctive genetic patterns in different regional stocks. By January 1936 they had joined in a systematic study of the evolutionary relationships between various regional populations. Dobzhansky wrote Demerec that they had "gone crazy with the geography of inversions in pseudoobscura and working on this whole days— he [Sturtevant] with crosses and myself with the microscope." Just a month later he reported in high excitement that they were "constructing *phylogenies* of these strains, believe it or not." No one was more surprised than Dobzhansky: "This is the first time in my life I believe in constructing phylogenies, and I have to eat some of my previous words in this connection. But the thing is so interesting that both Sttt and myself are in a state of continuous excitement equal to which we did not experience for a long time." From overlapping inversions they were able to infer the order in which inversions had appeared: that is,

30. Dobzhansky, oral history, pp. 358–66, 403–4; and Provine, "Origins," pp. 37–40.

31. Dobzhansky to Stern, 14 Jan., 21 Sept. 1935, CS; and Dobzhansky to Demerec, 21 Apr. 1935, MD.

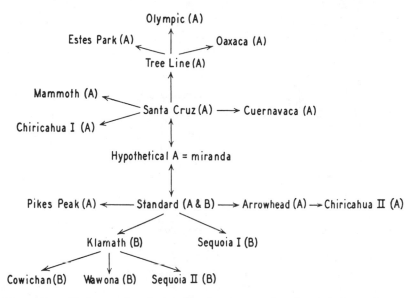

Figure 8.3 Phylogenetic relationships between regional populations of *D. psedoobscura*, inferred from overlapping chromosomal inversions. From Dobzhansky and Sturtevant, "Inversions in the chromosomes of *D. pseudoobscura*," p. 51.

the sequence in which local varieties had evolved (figures 8.3, 8.4). Hybrid sterility was left behind forever.[32]

Dobzhansky and Sturtevant had unexpectedly hit upon a natural source of varied genetic material that equaled in its rich diversity what the mapping project had produced twenty years earlier in *D. melanogaster*. The varied ecosystems of western North America were a natural breeder reactor in which isolation in forest and canyon microenvironments produced and preserved an unusual richness of genetically diverse local stocks. (The Mexican transition zone between temperate and tropical regions was especially fruitful of new varieties and species.) Because of its cosmopolitan habits, *melanogaster* displayed no comparable geographical variety; it became a breeder reactor only in

32. Dobzhansky to Demerec, 26 Jan., 6, 17 Feb., 12 Dec. 1936, MD; Provine, "Origins," pp. 30–42; Dobzhansky, oral history, pp. 404–5; A. H. Sturtevant and Th. Dobzhansky, "Inversions in the third chromosome of wild races of *Drosophila pseudoobscura*, and their use in the study of the history of the species," *Proc. Nat. Acad. Sci.* 22 (1936): 448–50; and Dobzhansky and Sturtevant, "Inversions in the chromosomes of *Drosophila pseudoobscura*," *Genetics* 23 (1938): 28–64.

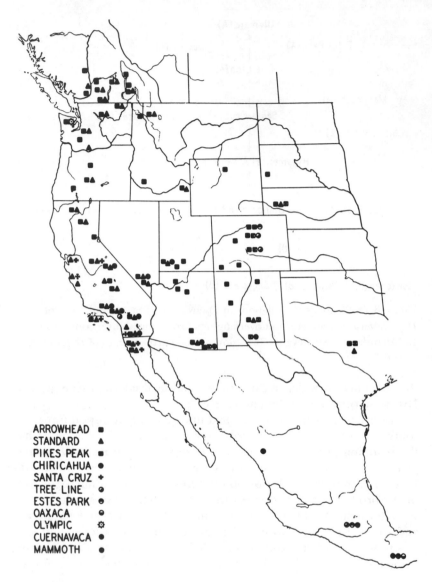

Figure 8.4   Geographical map of regional populations of *D. pseudoobscura*. From Dobzhansky and Sturtevant, "Inversions in the chromosomes of *D. pseudoobscura*," p. 53.

the peculiar ecology of a large-scale genetics lab. Not so with the forest-dwelling populations of *pseudoobscura*; they entered the lab with a richness of genetic variability that had accumulated over innumerable generations and thousands of years. The microhabitats of mountain forests were, for Dobzhansky, the equivalent to the fly group's mapping project: the generator of an endless wealth of experimental material, and a rare opportunity to study experimentally the process of variation and evolution in nature.

What enabled Dobzhansky and Sturtevant to detect these regional genetic patterns was the larger scale on which they were working by 1935–36. Regional differences are large-scale phenomena, visible only to experimenters who work on copious numbers of flies collected from many places. Just as the phenomenon of mutation did not become visible to Morgan until his breeding experiments crossed a certain threshold of scale, so, too, Dobzhansky and Sturtevant hit a critical threshold of scale after which they began to perceive the distinctive genetic patterns of regional populations of *pseudoobscura*. Scaling up made this natural genetic resource visible for the first time and thus accessible to laboratory experiment. Dobzhansky did not make this move consciously; he and Sturtevant became aware of the unexpected consequences of working on a larger scale only after the critical threshold had been crossed and meaningful patterns began to appear before their eyes.

Two seemingly innocent changes in working methods enabled Dobzhansky to work on this larger scale. First, as he pursued the search for hybrids, his collecting expeditions became ever more elaborate and far-ranging, covering many different biogeographical regions. That meant more flies from more diverse locales. Second, the invention of salivary chromosome cytology gave him a tool for identifying inversions that was much faster and more convenient than classical genetic analysis. Inversions, duplications, and translocations that before had been laboriously worked out from linkage data could now be seen simply by inspecting cytological slides. Rapid genetic screening made it possible for Dobzhansky to work up the larger number of wild flies that he was collecting in the field. These simple changes in laboratory and field practice reinforced each other in a positive feedback loop: the more flies Dobzhansky could work up in the lab, the more he could collect in the field. Once the critical threshold of scale was crossed, *pseudoobscura* was quickly transformed from a tool of classical genetic analysis into a system for experimental evolutionary genetics.

Theophilus Painter's new method using giant salivary chromosomes, announced late in 1933, took the *Drosophila* world and the fly group by storm. Bridges, who had never dissected a *Drosophila* larva in his life (Dobzhansky showed him how), at once launched into systematic cytological mapping of *melanogaster*.[33] Dobzhansky and Sturtevant's work with *pseudoobscura* was also transformed. Sturtevant was spared what remained of a "back-breaking" genetic analysis of a large inversion, the occurrence of which he thought might be what had separated races A and B. Dobzhansky and a postdoctoral fellow, C. C. Tan, similarly applied the new technique to their work on hybrids of *miranda* and *pseudoobscura*, revealing to their surprise and delight many small inversions and six translocations that they had failed to detect with standard genetic analysis! "Here, at last," Dobzhansky rejoiced, "you have real differences in chromosomes of different species."[34] Soon, he would be applying the new salivary assay to regional populations on a scale that would have been quite impracticable with classical genetic methods.

Meanwhile, Dobzhansky's annual summer collecting expeditions were taking him further and further afield. In 1934 he collected in Alaska, British Columbia and, as an afterthought, Idaho, where expert entomologists, including Sturtevant, had assured him pseudoobscuras were not to be found. (They were, if you knew their haunts and when they were about.) In 1935 he collected in Colorado, Arizona, and New Mexico and, again as an afterthought, in the vicinity of Mexico City. The Mexican populations, it turned out, were especially rich in inversions, and Dobzhansky regretted that he had not gathered more there.[35] Never before had drosophilists had such a wealth of wild material from such widely scattered regions of the Pacific slope and Great Basin. It was in these stocks, collected in the summer and fall of 1935, that Dobzhansky and Sturtevant first perceived distinct regional patterns—and the prospect of doing a real evolutionary genetics.

33. Dobzhansky, oral history, pp. 327–29; T. H. Morgan, "Calvin Blackman Bridges," *Biog. Mem. Nat. Acad. Sci.* 22 (1941): 31–48, on pp. 38–39; Weaver diary, 24 Apr. 1934, and Hanson diary, 14 Nov. 1934, RF 1.1 205D 7.86; and Demerec to Dobzhansky, 9 Jan. 1935, MD. Inversions, for example, appeared under the microscope as loops that formed when normal and inverted segments of sister chromosomes failed to pair.

34. Dobzhansky to Stern, 14 Jan. 1935 (see also 21 Sept. 1935), CS; and Th. Dobzhansky and C. C. Tan, "Studies on hybrid sterility, III: A comparison of the gene arrangement in two species, *Drosophila pseudoobscura* and *Drosophila miranda*," *Z. Indukt. Abstamm.* 72 (1936): 88–114. Dobhzansky and Hans Bauer were doing a similar study of hybrids of *D. athabasca* and *D. azteca*.

35. Dobzhansky, oral history, pp. 370–79, 380–83, 419–34.

But if the salivary assay and wider collecting created the potential for an evolutionary genetics, there remains the question why Dobzhansky and Sturtevant perceived evolutionary genetics as an opportunity that was so important that it just had to be seized. Although we have no direct evidence of their changing perceptions, this aspect of the story is in some ways the least problematic because it was overdetermined. Dobzhansky had long been interested in the genetic structure and dynamics of wild populations and had been following Romashov and Dubinin's recent work on the surprisingly heavy load of lethals and other mutations present in wild drosophilas. Sturtevant, similarly, had long been interested in evolutionary relations between species. Both were prepared to recognize the evolutionary opportunity in Dobzhansky's *pseudoobscura* stocks. Their minds were prepared, and their eyes open; that is why they were so quick to see their chance, drop what they were doing, and exploit the new system that unexpectedly fell into their laps.

The limiting factor in change, I believe, was not perceptions but experimental practicalities. The potential for an evolutionary genetics was created by the scaling up of Dobzhansky's fieldwork combined with the new salivary assay. This new system of field and laboratory work gave drosophilists access to the genetic richness of natural populations. It transformed *D. pseudoobscura* from a minor variant of the standard fly to a new system for experimenting on the process of variation and speciation in nature. Evolutionary genetics became doable, just as genetic mapping had twenty five years earlier. As Morgan's scaled-up selection experiments revealed the potential of *melanogaster,* so did Dobzhansky's scaled-up field collecting reveal the potential of wild *pseudoobscura*—another breeder reactor. But whereas Morgan was led away from experimental evolution to chromosomal mechanics, Dobzhansky was led from classical genetics back to experimental evolution.

Partners in Obscurology

Dobzhansky and Sturtevant's collaborative analysis of regional populations of *pseudoobscura* was extraordinarily intense and productive, rather like Beadle and Ephrussi's collaboration on transplantation (though shorter-lived). By the end of 1936 Sturtevant and Dobzhansky had analyzed four hundred strains and identified seventeen related populations. "I have definitely left the track of the classical Drosophila-forschung," Dobzhansky wrote Stern, "and [am] becoming more and more engaged in subjects that are somewhat abhorrent (or at least not very interesting) for the drosophila fellows here. Fortunately Sturtevant (though sometimes slightly apologetically) is going in the same direction, and in fact at the present we are doing most things jointly."[36]

The heady experience of their work on inversions and phylogenies encouraged Sturtevant and Dobzhansky to expand their collaboration into a more ambitious and focused program of field research on wild populations. In March 1936 Sturtevant sent Sewall Wright the outline of a comprehensive program, hoping it would entice Wright to join them and contribute his formidable skills in theoretical population dynamics. Although it was written and signed by Sturtevant, this program was very much a joint work. Sturtevant used "we" not "I" and recounted everything that he and Dobzhansky had done, separately and together, without distinguishing what belonged to whom.[37] Such a comprehensive collaboration went a good deal beyond what had been customary in the fly group, and in hindsight it went too far.

The trouble was that the work on regional populations was too close to Sturtevant's and Dobzhansky's deepest personal interests. It brought them into direct competition and afforded no room for division of labor, no way of keeping distance. In previous collaborations their individual contributions had been complementary, not competitive. Dobzhansky would not have wanted to do the genetic analysis of the A-B inversion, nor Sturtevant the analysis of sterility genes. Sex ratios and maternal effects did not threaten vital personal interests. But when Sturtevant and Dobzhansky began to map regional populations and construct phylogenies, they converged on a problem about which both cared deeply and about which they disagreed fundamentally. Their idea of a big team project on the genetics of natural populations put them on a collision course.

It was already clear in early 1936 that Sturtevant and Dobzhansky had fundamentally different conceptions of what evolutionary genetics was all about. Sturtevant was interested in traditional questions of systematics and phylogeny—the historical relations *between* species. For Dobzhansky, in contrast, the fundamental issue was the process of speciation *within* species. They also had quite different conceptions of what a species was. Dobzhansky, following Chetverikov, viewed species as collections of different local populations. Sturtevant, in contrast, had a rigidly typological view of species. As Ernst Mayr later observed,

36. Dobzhansky to Stern, 26 Feb. 1936, CS.

37. Sturtevant to S. Wright, 18 Mar. 1936, SW; Sturtevant, "Draft and prospect of the Drosophila pseudoobscura analysis," SW; and Provine, "Origins," pp. 42–45.

"he almost acted as if he considered every species genetically homo-zygous."[38] In other words, Sturtevant treated wild flies as if they were standard, domesticated melanogasters. Dobzhansky was always aware of the difference, even in the lab.

The products of their experiments were also quite different. Stur-tevant's genetic analysis would produce chromosome maps and, ulti-mately, phylogenetic trees. Dobzhansky's work on natural populations would produce biogeographical maps of distinct local populations and, in time, charts or equations describing the dynamics of local popula-tions as they were subject to changes in physical environment or in competition from other populations. These different visual representa-tions embody two quite incongruent modes of experimental practice.

Differences of purpose were also rooted in Sturtevant and Dob-zhansky's divergent styles of working. Sturtevant put a high value on rigorously quantitative genetic analysis; Dobzhansky preferred quicker and more qualitative biometric methods.[39] Sturtevant liked to analyze a few cases with great precision; Dobzhansky preferred a quick-and-dirty survey of a wide range of material. Dobzhansky was a romantic, willing to make bold leaps of interpretation, while Sturtevant was cau-tious and mistrusted generalizations. These differences in aim and work style were complementary when the two were working on differ-ent lines, but they became bones of contention when Sturtevant and Dobzhansky set to work on the same problem. Personal conflicts also surfaced as the sensational results with *pseudoobscura* catapulted Dob-zhansky from junior partner to Sturtevant's equal in the fly group.

The Dobzhansky-Sturtevant Rift

The first intimation of a rift between Dobzhansky and Sturtevant ap-peared in May 1936, barely two months after they had hatched their grand collaborative project. The story, in brief, is this. In May 1936 Dobzhansky was unexpectedly offered a professorship of genetics at the University of Texas. It was an extremely flattering offer, offering the prospect of star colleagues, lots of graduate students and postdocs, a hard-driving work style, and ample money to expand the work on *pseudoobscura*—and, of course, independence. Not wishing to lose Dob-

38. E. Mayr to Provine, 19 Nov. 1979, WP.

39. For example, Sturtevant identified inversions by measuring linkage in elaborately constructed stocks; Dobzhansky inferred the presence of sterility genes from measuring the size of testes in hybrids.

zhansky, Morgan promised to promote him to full professor. At the same time Demerec, ever alert to opportunity, was dangling the prospect of a position at Cold Spring Harbor. What to do? Dobzhansky was unprepared to deal with the practical problems of stardom and turned for advice to Sturtevant, his trusted friend. Although Sturtevant refused to tell him what to do, he made it clear that he thought Dobzhansky should accept the Texas offer. Demerec was also urging him to "cut loose from Morgan," so Dobzhansky decided to accept. But might that prevent him from accepting an offer from Cold Spring Harbor? Should he perhaps resign from Caltech, decline Texas, and wait for Demerec to come through? Dobzhansky again consulted Sturtevant, who told him to do no such thing, when he could easily move on from Texas if he wanted to. That seemed reasonable, but Natasha thought that it was silly to move twice, so Dobzhansky decided to stay and wrote Patterson declining his offer.[40] Morgan was delighted, but not Sturtevant: "I told him, 'We stay,'" Dobzhansky recalled. "His face fell. It was quite obvious he didn't want this. That was a terrific shock." Sturtevant, his best friend, wished he would leave! Dobzhansky rushed off a second letter to Patterson accepting the post, but it was too late: the position had been canceled.[41]

This episode was not an immediately fatal blow to Sturtevant and Dobzhansky's friendship. They continued to collaborate, but warmth and trust gradually gave way to a cool and wary standoff. Sturtevant vacated the laboratory room he had shared with Dobzhansky since 1928, and shoptalk became less frequent and more strained. By the end of 1938 they were working on quite separate lines. Sturtevant continued to work on the comparative genetics and evolutionary relations of different species of *Drosophila*. Dobzhansky confined himself to the dynamics of local populations of *pseudoobscura* and the mechanism of speciation. Through a kind of internal isolation Dobzhansky and Sturtevant became different subspecies of drosophilist, inhabiting the same laboratory space but no longer interfertile—like the A and B races of *pseudoobscura*. By the time Dobzhansky left the fly group in 1940 he was sure that Sturtevant was his enemy. What was going on?

Various explanations have been proposed for the rift that divided Dobzhansky and Sturtevant, most of which emphasize their different

40. Dobzhansky, oral history, pp. 384–88; Provine, "Origins," pp. 45–46; Dobzhansky to Demerec, 7, 20 May 1936, MD; Demerec to Dobzhansky, 11 May 1936, 4 Jan. 1938, MD; and Demerec to Dobzhansky, 18 May 1936, TD.

41. Dobzhansky, oral history, p. 397.

temperaments and work styles.[42] However, the two had had tiffs before without seriously disrupting their friendship and collaboration. In 1933 or 1934, for example, Dobzhansky was stunned when Sturtevant suddenly turned on him for making a critical remark about Morgan.[43] Yet their most intimate partnership was to come. So, why was the potential for serious conflict so much greater in 1936–37? The main reason, I believe, was the greater intensity of Dobzhansky and Sturtevant's collaboration on regional mapping and the construction of phylogenies. Evolutionary genetics was too important to both to be shared. Total collaboration left no room for a division of labor, which previously had always defused potential rivalries. As the structure of experimental work changed, the group's moral economy ceased to work as it once had.

Sturtevant and Dobzhansky worked in different places, on different kinds of material, to different ends. This divergence in experimental practice is apparent in Dobzhansky's collecting practices, which focused on smaller and smaller geographical areas. In 1936 he collected in the island mountain ranges of Death Valley and the Mojave Desert, seeking distinctive genetic types in isolated populations. In 1937 he collected in the separate canyons and valleys of each range, after realizing that there was considerable genetic variation even among very localized populations. (Before then he simply pooled all the flies collected in an entire range.) This experience in the recesses of the Panamint Mountains was what finally caused Dobzhansky to abandon the last vestiges of the typological view of species, which had informed his work with Sturtevant and to which Sturtevant remained wedded.[44]

Localized collecting was of little use for Sturtevant's interest in phylogeny, and he grew critical and disdainful of Dobzhansky's methods. "Sturtevant did everything in his power to discourage this work," Dobzhansky later recalled. "He argued that the technique of detection of this concealed variability was not valid, and that the thing has no partic-

42. Provine, "Origins," pp. 47–53.

43. Dobzhansky, oral history, pp. 273–74; Dobzhansky to Mayr, 29 Mar. 1975, TD; Provine, "Origins," pp. 30–32; and Morgan to O. Mohr, 16 Mar. 1922, THM-APS.

44. Dobzhansky, oral history, pp. 406–7, 416–18, 437–39; Mayr to Provine, 1 Aug., 19 Nov. 1979, 7 Feb. 1980, WP; Dobzhansky to J. C. Merriam, 9 Sept. 1938 and "Concerning proposed studies on the genetics of natural populations," CIW; Th. Dobzhansky and M. L Queal, "Genetics of natural populations, I: Chromosome variation in populations of *Drosophila pseudoobscura* in-

ular interest anyway."[45] Sturtevant remained an eager consumer of wild material that Dobzhansky collected, but now in a way that was quite different from Dobzhansky's own uses of it. Leaving *pseudoobscura* alone, Sturtevant confined himself to the many other species that wandered into Dobzhansky's traps. An element of reciprocity remained; in return for the wild species Sturtevant cared for the shipments that Dobzhansky sent back from the field. Dobzhansky was glad to share but disdained Sturtevant's conservative tastes. Afield in Mexico in 1938, he wrote Demerec of Sturtevant's delight with what had been sent so far: "Six new species of the *hydei* group alone, and . . . the total number of new species is almost a dozen! But, of course, what is important are not the new species but the old ones—*pseudoobscura* and *azteca*."[46] Dobzhansky was no longer collecting for one unified project, as he had been in 1933–36, but for two individual projects and two distinctly different modes of experimental practice.

The Fly Group Succession

Heightened social tensions in the fly group in the late 1930s may also have intensified the rift between Sturtevant and Dobzhansky. The cause of these intragroup tensions was Morgan's retirement: more precisely, his reluctance to retire.[47] The years between 1936, when he turned seventy, and 1940, when he more or less retired, were years of increasing uncertainty and tension, and Sturtevant, as Morgan's heir apparent, took the brunt. Sturtevant probably expected to be asked; he would never say so, of course, but Demerec thought that he would not refuse Morgan's chair if it were offered to him.[48] It would have been just reward for a lifetime of self-denial. But much as Morgan had failed

habiting isolated mountain ranges," *Genetics* 22 (1938): 239–51; Dobzhansky and Queal, "Genetics of natural populations, II: Genic variation in populations of *Drosophila pseudoobscura* inhabiting isolated mountain ranges," ibid., pp. 463–84.

45. Dobzhansky, oral history pp. 404–15, on p. 415.

46. Dobzhansky to Demerec, 24 Mar. 1938, MD; and Dobzhansky to Dunn, 2 Apr. 1938, LCD.

47. Morgan agreed when he came to Caltech in 1928 that he would retire in five years to make way for a younger man or men: Morgan to Millikan, n.d. [22 May 1933], THM-CIT 1.

48. Demerec to E. M. East, 20 July 1935, MD.

to delegate control of the genetics department to Sturtevant in 1928, in the late 1930s he put off the expected succession to his own chair. It was not just a desire to hold on: Morgan had reason to fear that the CIW would terminate its support when he retired, as we have seen. He also worried that Sturtevant was too indecisive to lead the biology division—as he warned Caltech's president Robert Millikan.[49] A mood of anticipation turned into nagging uncertainty as months and years went by with no sign from Morgan what he meant to do. Sturtevant became increasingly unhappy and on edge.

Uncertainty over the succession upset the fragile balance of personal authority among the members of the group. Personal ambitions were no longer held in check by deference to Morgan, and personal tiffs could easily be magnified into institutional rivalries. That, I believe, is what happened to Sturtevant and Dobzhansky. As a result of their work together on *D. pseudoobscura*, Dobzhansky became Sturtevant's equal and potentially his rival. I do not wish to claim too much. There is no evidence that Dobzhansky ever had his eye on Morgan's chair, or that Morgan ever thought of him as his successor. But I do think that as Dobzhansky's prestige and status grew, Sturtevant began to feel a certain loss of elbow room, a diminished psychosocial space. Sturtevant was accustomed to being the fly group's prima donna. Dobzhansky had similar tendencies, and as he graduated from junior to senior member of the group he too began to act like a prima donna, in a group where there was room for but one.

For Sturtevant, May was the cruelest month. Dobzhansky's promotion to full professor in May 1936 made him formally Sturtevant's equal. The offer from Texas and the talked-of offer from Cold Spring Harbor were plums that Sturtevant would have been honored to have. To top it all off, Dobzhansky was invited to give the prestigious Jessup Lectures on evolution at Columbia and to publish them as a book. Naturally, he chose to talk about *pseudoobscura*. It meant that he would be publishing under his name alone work that he and Sturtevant had done together. How could Sturtevant not have taken that hard? No drosophilist had devoted himself more to systematics and comparative genetics than had Sturtevant. And here was Dobzhansky getting the public acclaim. No wonder that Sturtevant was unable to hide his true feelings when caught by surprise by Dobzhansky's refusal of the Texas offer. Dobzhansky had become a rival, and Sturtevant would have been relieved, if also sad, to see him go.

49. Morgan to Millikan, [?] Apr. 1939, RAM 18; and Sturtevant to Millikan, 9 Apr. 1939, RAM 18.

Dobzhansky understood a little of his friend's emotional double bind, but Sturtevant's reticence left much room for misunderstanding. "Sturtevant knows no more than I do what will happen here after Morgan retires," he reported to Demerec in May, "and does not know when that is going to happen. He is not very hapy [*sic*] himself about the whole picture." A few months later, however, he was less sure that Sturtevant had been frank and candid with him.[50] He assumed Sturtevant's friendship for him was cooling, but did not see how Sturtevant's ambiguous situation with regard to the succession made him sensitive to potential rivals. Disruption of the group's structure of authority gave their disagreement over experimental practice a personal and institutional edge.

Conflict may even have impelled Dobzhansky to evolve experimental practices that were different from Sturtevant's, as a way of keeping distance. I am not saying that Dobzhansky invented a new mode of practice to get away from Sturtevant. But would he have diverged so far from traditional practices had he wanted to preserve their collaboration? We may speculate that personal disaffection amplified the difference between modes of practice that were already pulling apart. Intragroup conflict, by isolating Dobzhansky from the traditional practices of his American colleagues, cleared the way for the rapid evolution of a new species of experimental practice.

A Hybrid Practice: Field

In effect, Dobzhansky escaped from traditional laboratory practices by breaking down the threshold between laboratory and field. It was the equal combination of field and laboratory methods that made his system novel and productive, and thus competitive with standard systems. Before, when drosophilists brought wild flies indoors they handled them as they did *melanogaster*, working with a few specimens that were stripped of their natural diversity and made into standard laboratory flies. Dobzhansky, in contrast, brought indoors thousands of wild flies from scores of local sites. He carefully preserved their natural diversity—it was the object of his study. Domestication of *melanogaster* also made drosophilists into domesticated, indoors creatures. Dobzhansky and the new generation of evolutionary geneticists moved more freely across the threshold between laboratory and field.

50. Dobzhansky to Demerec, 20 May, 25 Nov. 1936; see also Sturtevant to Demerec, 26 May 1936; all in MD.

Figure 8.5 Theodosius and Natasha Dobzhansky on a field trip in Mexico, 1938. Courtesy of Sophie Dobzhansky Coe.

For Dobzhansky, the ability to move across that threshold was what he liked most about his new mode of experiment. His addiction to camping and his delight in wild and exotic landscapes were legendary. He was teased about it, but it was true:

> Since my wife and myself were both addicted to excursions, camping trips and so on, there was our chance, and more than one friend and colleague since have suspected that I started to work on Drosophila pseudoobscura . . . as a thin rationalization of this desire to travel!! The story was going around among geneticists that for other obscure reasons, Drosophila pseudoobscura occurs in the United States chiefly in national parks and in Mexico, chiefly near the ancient ruins. Anyway, I think I can fully and frankly admit that travelling, collecting, was something which attracted me a great deal.[51]

It is not surprising that he chose not to domesticate free-living creatures to the laboratory but rather to seek them out in their wild haunts.

Dobzhansky had always been an ardent camper and field collector, and he responded to landscapes in an intensely emotional, visceral way: the wilder and more romantic the better he liked them. As a young man he made a romantic excursion and collecting expedition in the Caucasus Mountains and Turkestan. In the 1920s he turned a government survey of Turkoman horse husbandry into a genetic field investigation.[52] These early experiences gave him a permanent yen for travel and field collecting. It did not take him long, after moving to California, to begin to explore the spectacular scenery of the North American West: high sierra, deserts, evergreen forests were all within relatively easy reach of Pasadena by automobile. The vast Alaskan outback delighted him: "All those snow peaks and forests," he marveled, "stupendous and indescribable."[53] The only thing Dobzhansky missed when he finally left Caltech was the Western landscape, and that he felt keenly. Whenever a letter from Schultz arrived for him at Columbia, it would evoke vivid images of "the mountains out of the window, the live oaks on the campus. . . . [W]hat a joke of fate that you all to whom mountains and deserts are relatively or quite unimportant have the privilege of seeing them, and I who loves them so much had to leave them behind!"[54]

51. Dobzhansky, oral history, p. 368.

52. Provine, "Origins," pp. 7–13.

53. Dobzhansky to Stern, 14 Jan. 1935, CS.

54. Dobzhansky to Schultz, 21 Mar. (quotation), 19 Jan. 1941, JS; and Dobzhansky, oral history, pp. 290–91.

Dobzhansky's collecting expeditions evolved gradually out of his camping excursions, each one becoming more elaborate and far flung: the Pacific Northwest (1933); Alaska and the Snake River basin (1934); the Four Corners region of Colorado and New Mexico (1935); Death Valley and the Mojave Desert (1936–37); Mexico and Guatemala (1938); Brazil, India, the South Pacific, and so on.[55] Dobzhansky may not have anticipated that these excursions would become such a large part of his work; in January 1935 he told Demerec that his forthcoming trip to Mexico would probably be his last.[56] In fact, continual collecting of wild material only became indispensable when he gave up hybrid sterility and phylogenies to study the genetics of local populations. Then fieldwork set the pace and direction of Dobzhansky's experimental life. Continual fieldwork was what demarcated his new mode of evolutionary genetics from more traditional modes.

It was not certain at first that Dobzhansky would be able to find enough wild pseudoobscuras for large-scale experiments. Unlike *D. melanogaster*, that hanger-on of humankind, *D. pseudoobscura* was a shy beast, feeding on bleeding tree sap and avoiding the traps of natural historians. The customary way of trapping wild flies, by setting out baited tubes for a few days, did not work for these elusive forest creatures. On his first collecting trip, however, Dobzhansky discovered that wild pseudoobscuras were active only around dawn and dusk and not during the day, when drosophilists, being creatures of diurnal habit, were afield. Sturtevant thought Dobzhansky had gone crazy when he told him: decades of observing drosophilas in the laboratory had given no hint of a diurnal habit. (In laboratories, of course, there were no extremes of temperature and dessication to avoid.) In any case, *D. pseudoobscura*'s habits fit better with the routines of campers than those of daytime entomologists. Dobzhansky had only to park his car for the night, set out traps, make camp, and wait until nightfall, by which time he had dozens or hundreds of little captives. In this way he could collect many samples quickly from a large number of locales.[57]

55. Dobzhansky, oral history, pp. 419–34; and *The Roving Naturalist: Travel Letters of Theodosius Dobzhansky*, ed. Bentley Glass (Philadelphia: American Philosophical Society, 1980).

56. Dobzhansky to Demerec, 19 Jan. 1935 (see also 10 Dec. 1934), MD; and Dunn to Dobzhansky, 24 Apr. 1938, TD.

57. Dobzhansky, oral history, pp. 330–39; and Sturtevant to Demerec, 27 Aug. 1927, MD.

The organization of Dobzhansky's field excursions varied a good deal but tended to become more elaborate as he went further afield. His trip to the redwood forests of Northern California in 1933 was more or less a long weekend of camping. The survey of the Southwest in 1935 kept him afield for several weeks. Collecting in the mountain ranges of Death Valley was somewhat more elaborate: Dobzhansky did much of it himself in the course of a summer, scheduling a series of short forays to half a dozen ranges. The Alaskan and Mexican trips, in contrast, were real expeditions. They required getting grants (the Rockefeller Foundation financed his trip to Alaska, and the Carnegie Institution the ones to Mexico).[58] And they involved formal arrangements for travel, which sometimes went awry, as when a dock strike left him stranded halfway to Alaska. He also needed local support networks. In Mexico, for example, the Carnegie Institution's field stations and officials of the United Fruit Company were most helpful.

The logistics of collecting were not unduly complicated but had to be carefully attended to. Usually Dobzhansky went afield with Natasha or a friend—the Hungarian-born cytologist Peo Koller was with him in the Panamint Mountains—or couples like L. C. Dunn and his wife.[59] There were the usual details of provisioning and arranging for cabins at regular intervals (not for the beds but the baths). Getting up and down the rugged canyons of the Panamints required two donkeys to carry supplies. Most important, Dobzhansky had to arrange accommodation for his captives. He had to make sure that he had a sufficient supply of culture vials and fly food and that he was never too far from a post office.

Getting cultures back to the lab was a critical step: pseudoobscuras were sensitive to heat and could easily end up dead on arrival if the mails were delayed. The captives in their glass cells were mailed as soon as possible to Caltech, where Sturtevant or Dobzhansky's assistants were waiting to transfer them into the safer conditions of laboratory incubators. Dobzhansky could have organized a traveling field laboratory for routine screening, but the public mails were more convenient. There were few places in North America that were not served by the commu-

58. Dobzhansky to Hanson, 17 Jan., 15 Feb. 1934, 8 Feb. 1935, RF 1.1. 205D 7.86; Dobzhansky to Merriam, 17 Feb. 1937 and "The general purpose of the proposed work," CIW file Morgan; and Dobzhansky to Merriam, 9 Sept. 1938—and much else, all in CIW.

59. Dobzhansky to Dunn, 10 Mar. 1940, 4, 15 June, 9, 17 Aug. 1941, 4 Aug. 1942, LCD; and P. Koller to Muller, 27 May, HJM-C.

Figure 8.6 Dobzhansky trapping wild *Drosophila* in the field, circa 1950. Courtesy of Sophie Dobzhansky Coe.

nication infrastructure of modern civilization.[60] The Guatamalan highlands and the mountains of Death Valley were not the tamed suburban nature of Woods Hole, but neither were they the empty steppes of Central Asia.

Field collecting involves conventionalized experimental routines no less than does laboratory experiment. One needs to know where to look for creatures with different habits and microhabitats, in what seasons, at what time of day, in what weather, with what traps and bait, and so on. It is striking how long the business of field collecting remained a craft tradition, tacit knowledge that could not be learned from books but only from field experience. It was not the custom for field drosophilists to describe their working methods in print, as laboratory experimenters did. When Warren Spencer agreed in the late 1940s to write a chapter on fieldwork for a *Drosophila* handbook, he had to gather his material by talking to experienced collectors.[61]

60. Dobzhansky to Demerec, 2, 17 Feb., 24 Mar., 2 Apr. 1938, MD; and Dobzhansky, "Collecting, transporting, and shipping wild species of *Drosophila*," *DIS* 6 (1935): 28–29.

61. W. Spencer, "Collection and laboratory culture," in M. Demerec, ed., *Biology of Drosophila* (New York: Wiley, 1950), pp. 535–87, on p. 537.

Dobzhansky was never a field biologist in the same way as natural historians or field ecologists were, whose principle workplaces were remote field camps. The difference was apparent when Dobzhansky spent a summer with the botanists Herman Spoehr and Jens Clausen at the Carnegie Institution's field camp in the California Sierra. Dobzhansky had a wonderful time, but Spoehr and the Clausens had to make sure he did not tangle with snakes or wildfire or get lost: "He is an extremely impractical person so far as field work is concerned," Spoehr noted. "We had to help him out of many spots and Clausen is there now largely for that purpose." Dobzhansky had noticed that Clausen "treated us with . . . parental care" but did not quite understand why.[62] From the point of view of real field biologists Dobzhansky was a man with one foot still in the laboratory. "Dobzhansky is just emerging out of the milkbottle (T. H. Morgan) stage of Drosophila genetics," Spoehr observed. "[H]e is regarded as a bit of a radical among the conventional geneticists. But his outlook is still quite conventional as compared with the Clausen group."[63] Precisely the point: Dobzhansky was a hybrid. In the field he seemed to be a laboratory geneticist, in the lab, a natural historian. He was not an ecologist or field naturalist, nor was he a "milk-bottle" gene mapper. Fieldwork was one half of a mode of practice that united field and laboratory in equal measure. That was precisely what made it so productive and attractive to drosophilists.

A Hybrid Practice: The Laboratory

Fieldwork was just the first stage of Dobzhansky's experimental population genetics. The real work began when the bottles of wild flies arrived in the laboratory. Each fly was individually cultured, and the chromosomes of its offspring were screened for visible inversions and translocations. Then the genetic search began for recessive mutations, lethals, and so forth, by means of systematic crossings with standardized, marked laboratory strains. Those genetic data were the raw material for identifying distinctive local populations and mapping their biogeographical boundaries. Material collected in a few weeks afield kept

62. H. A. Spoehr to W. M. Gilbert, 14 Sept. 1945; and Dobzhansky to Spoehr, 21 Sept. 1945; both in CIW file Dobzhansky.

63. Spoehr to Gilbert, 8 Oct. (quotation), 9 Nov. 1945, and Clausen memo, 8 Nov. 1945, all in CIW file Dobzhansky; and Dobzhansky to Dunn, n.d. [early May 1938], TD.

Figure 8.7   Dobzhansky at
work in the laboratory of
the Carnegie Institution's
field station at Mather, California, circa 1950. Cour-
tesy of Sophie Dobzhansky
Coe.

Dobzhansky and his assistants busy for many months. The work could
not be put off: females arrived in the laboratory pregnant, and the
salivary technique required that larvae be dissected before they got too
old. The pace of work after one of Dobzhansky's excursions was
frenzied.

Dobzhansky's letters give a vivid picture of the sheer bulk of the
material that poured in. The flies collected in a three-day trip to the
Providence mountains in Death Valley covered half of his large work-
table, and there were nine more trips planned. With four mountain
ranges visited, Dobzhansky was screening the chromosomes of two to
three hundred flies from each site and analyzing a hundred each of
chromosomes 2 and 3 (and doing it all himself, since his assistant was
on vacation).[64] After six trips to eleven mountain ranges, the work had
become monumental:

> The analysis of this mass of flies requires, we have counted, about 8000
> bottles. . . . Four persons (my assistant, two students working on a tempo-
> rary basis, and myself) are putting in full time in this business. Some

64. Dobzhansky to Demerec, 10, 28 May 1937, MD; and Dobzhansky to
Stern, 5 Feb., 16 Sept. 1937, CS.

evenings the work goes on til 11 p.m. Somehow I am an optimist in this case, and believe and hope that the results will justify all this expenditure of energy, time, and money (not much money involved, though).[65]

He was lucky the weather had been cool; otherwise the cultures that were spread out on every table would have had to be moved to cool rooms.

The seasonal character of field collecting had certain inconveniences. Dobzhansky was not used to working summers in Pasadena, when everyone else was having a good time at Woods Hole. He felt left behind. "I am utterly alone," he complained to Demerec. "Sttt has not been kind enough to write me a single letter the whole summer and I have no idea of what is going on in the outer world."[66] But he knew it was an unavoidable cost of doing business in the new way: "Having tied myself up with population problems," he admitted, "I have lost one of the essential advantages of a Drosophila geneticist, namely the independence from the time of the year. Now I have to sit and work in the summer, and take my vacation in winter if I want and can."[67]

It was not just the pace and scheduling of laboratory work that was changed by the invasion of wild pseudoobscuras; so, too, were its standards of quality. With so many flies to screen, compromises had to be made between scale and precision. For example, the acetocarmine preparations of salivary chromosomes were quick and easy to do but did not produce a permanent record, nor were they, perhaps, so impressive to classical cytologists as neat balsam slides. Working up thousands of wild flies also required a streamlined genetic analysis. Dobzhansky could not afford to examine flies closely for visible mutants, even if he had wanted to. He was not a good mutation finder, he cheerfully admitted, and adopted methods of analyzing for concealed genetic variability of which traditional geneticists like Sturtevant did not approve.[68]

To laboratory drosophilists, especially cytologists, Dobzhansky seemed to be a sloppy worker whose results could not always be trusted

65. Dobzhansky to Demerec, 25 July 1937 (quotation), see also 23 Apr., 14 May 1938, all in MD; and Peo Koller to Muller, 13 July 1938, HJM-C.

66. Dobzhansky to Demerec, 12 Sept. 1937, MD; see also Dobzhansky to Stern, 16 Sept 1937, CS.

67. Dobzhansky to Schultz, 5 Jan. 1939, JS.

68. Dobzhansky, oral history pp. 414–15.

to stand up.[69] Doubtless there was some truth in that. A critical cytologist revealed nine errors in Dobzhansky's work on *D. miranda,* and Franz Schrader trembled lest other cases be reopened. "The long and the short of it," he wrote, "is that Doby is a man with a brilliant mind who is constitutionally incapable of enduring the discipline of cytological labor and passing impassionate and detached judgement thereupon."[70] At the same time, the perception of "sloppy" workmanship was symptomatic of the fundamental changes in laboratory practices that followed the invasion of the wild drosophilas. It is hardly surprising that, in a community that put such a high value on puritanically quantitative analysis, Dobzhansky's shortcuts appeared to be a decline in standards. But Dobzhansky's standards were not lower than those of classical geneticists: just different. What was gained in a greater breadth, in Dobzhansky's genetic surveys of wild flies, was paid for by a loss in precision.

Such changes in the moral perception of different practices are probably quite common when exact methods from one discipline are applied on a larger scale in another one. The biophysicist William Astbury was regarded as somewhat deviant by X-ray crystallographers, for example, when he adapted their exacting technology to rough-and-ready measurement of the shape of protein fibers. Similarly, Linus Pauling had to exhort chemists not to be deterred from studying complex organic molecules by awe of physicists' exceedingly precise measurements of very simple inorganic crystals. For their purposes, chemists needed lots of data that did not need to be very precise. Such practices did not make chemists sloppy, just different from physicists.[71] So, too, with Dobzhansky's new mode of evolutionary genetics. Since his aim was to map the genetic structure not of chromosomes but of constantly changing local populations, precise cytogenetic methods were less appropriate than rough-and-ready survey.

The combination of delightful field excursions and binges of working up in the lab were very much to Dobzhansky's taste. He liked a fast pace of production, with lots of new data coming along regularly, and he was not happy doing more orthodox laboratory experiments. In 1938–39, for example, he slogged through a large project to measure

69. Provine, "Origins," pp. 48–50; see E. Novitski to Provine, 1 Dec. 1979; E. Mayr to Provine, 7 Feb. 1980; J. Bonner to Provine, 27 Dec. 1979; and J. A. Moore to Provine, 11 Sept. 1979; all in WP.

70. Schrader to Stern, 5 Dec. 1945, CS.

71. Robert E. Kohler, *Partners in Science: Foundations and Natural Scientists, 1900–1945* (Chicago: University of Chicago Press, 1991), pp. 334–36, 341–45.

the natural rate of mutation in a long-domesticated laboratory stock. The project involved thousands of cultures and stretched over nearly a year, and Dobzhansky was thoroughly fed up with it long before it was over. "Never," he complained to Schultz, "have I worked on a more dull, uninteresting, and tiresome problem than this."[72] He had the same experience with experiments on hybrids of *pseudoobscura* and other species; all fizzled, leaving Dobzhansky restless to get back to his wild flies. The cycle of field and laboratory offered a livelier pace, greater variety, and a greater likelihood of producing exciting new results.[73]

Sideshow: Sturtevant's Evolutionary Genetics

The success of Dobzhansky's new mode of experimental evolutionary genetics spurred Sturtevant to develop his own style of comparative genetics on an equally impressive scale. "Sttt is planning something of a big undertaking," Dobzhansky reported to Demerec in November 1936. "He wants to study comparative genetics of species of Drosophila of the world, an undertaking which would call for a budget of the order of $10,000 per year." Knowing Sturtevant's dislike of organized projects, however, Dobzhansky doubted that the project would ever get beyond "the purely conversational stage."[74] It was unusual behavior for Sturtevant, true, but then comparative genetics and systematics were subjects that had been unusually dear to his heart for almost twenty-five years. Sturtevant's plan was, in fact, an enormously expanded form of the big program of comparative genetics that he and Metz had hoped to mount in the early 1920s. The idea was to describe and map all the known mutant alleles of all the known species of *Drosophila,* identify homologous chromosomes, and from these homologies construct the phylogenetic relations between species.

Sturtevant did an immense amount of work on comparative genetics in the late 1930s and early 1940s, though little of it was ever published. By the end of 1939 he was cultivating no fewer than forty-two species. He had selected twenty-seven "standard" characters from all stages of the various species' life cycle and was busily comparing and tabulating differences. He was also collecting on a vast scale; more pre-

72. Dobzhansky to Schultz, 5 Jan. 1939, JS; Dobzhansky to Demerec, 10 Mar. 1939, MD; and Dobzhansky to Stern, 27 Jan. 1939, CS.

73. Dobzhansky to Demerec, 3, 10 Mar. 1939, MD.

74. Dobzhansky to Demerec, 25 Nov. 1936, MD.

cisely, he was collaborating with colleagues like John Patterson, Harrison Stalker, M. R. Wheeler, and Warren Spencer, who were collecting systematically and on a vast scale in the late 1930s and 1940s. (Sturtevant contributed his expert knowledge of systematics in return for access to new material from the field.) Three brief theoretical papers on species were probably meant to be a counterthrust to Dobzhansky's Jessup Lectures. The results of all his labors were surprisingly meager, however, and by the mid 1940s Sturtevant seems to have lost interest in his big project.[75] It was the Texas group and Spencer, not Sturtevant, who produced the natural and evolutionary histories of *Drosophila* species in the 1950s. One cannot be sure why Sturtevant lost interest, but one reason surely was that he simply did not love systematic, large-scale collecting in the field, as Dobzhansky, or Spencer or the Texas group clearly did.

This is not the place to go into further detail about Sturtevant's interesting but tributary work. The point here is that it epitomized the traditional practice of evolutionary genetics as Metz and Sturtevant had defined it in the 1910s; its aim was a comparative genetics of species, not the mechanism of speciation. Sturtevant treated wild species as if they were standard, domesticated types, and his rigidly typological view of species was congruent with his traditional experimental practices. He did need constantly fresh material for his phylogenetic work (even brief domestication could alter the natural genetic constitution of wild flies), but comparative genetic mapping also required that wild species be domesticated and transformed into standard flies. Sturtevant brought wild creatures into the laboratory to be assimilated into a domestic system of production. Fieldwork was an extension of domestic, laboratory customs.

Dobzhansky's experimental practice, in contrast, brought a bit of the wild indoors into the laboratory. He preserved the natural diversity of wild populations, because that diversity was the object of study. Traditional domestic practices were accommodated to the requirements

75. There is voluminous correspondence in AHS, 5 and 6. See also A. H. Sturtevant, "On the subdivision of the genus *Drosophila*," *Proc. Nat. Acad. Sci.* 25 (1939): 337–53; Sturtevant, *The Classification of the Genus Drosophila, with Descriptions of Nine New Species,* University of Texas publication no. 4213 (Austin, 1942), pp. 5–51; Sturtevant, "Essays on evolution," *Quar. Rev. Biol.* 12 (1937): 464–67; 13 (1938): 74–76, 333–35; and Sturtevant to Mohr, 30 Dec. 1939, 14 Sept. 1945, OM. Sturtevant told Mohr in 1939 that he was "rather surprised to find myself figuring out hypothetical phylogenies for the Drosophila species, and taking them more or less seriously—after all the uncomplimentary remarks I've published about such procedures."

of wild flies. Lab work became, to some degree, an extension of the field. Dobzhansky treasured the "natural" integrity of wild flies and held in scorn the artificiality of laboratory drosophilas, which were called "wild type" but which had long since been tamed.[76] He even came to scorn wild strains of *D. melanogaster* because even in their so-called natural state they were semidomesticated town and city dwellers, parasites of civilized humankind. Far better, he thought, to work with "a 'natural' species like pseudoobscura, and on mountains rather than on garbage cans and [in] store rooms."[77] In bringing wild nature indoors, Dobzhansky subverted traditional laboratory practices and earned for himself the reputation of being a little wild himself—like his new friends.

The differences between Dobzhansky's and Sturtevant's varieties of evolutionary genetics were rooted in the practical realities of experimental life. They were spatial, tactile, and visceral as much as cerebral and theoretical. Rival modes of practice were done in different places, had different rhythms and paces, evoked different feelings. They were different cultures of production.

Dobzhansky and the Fly Group, Final Years

Dobzhansky's last years in the fly group were extraordinarily productive but not very happy ones. His new mode of practice expanded too fast and in too many directions, outstripping local resources. It was too different from the traditional modes of practice that Sturtevant and others preferred. Dobzhansky came more and more to feel that he was the odd man out, that his colleagues disapproved of his interest in "such terrible things as taxonomy and morphology."[78] The more clearly his genetics of natural populations diverged from traditional comparative genetics, the more Dobzhansky felt that his colleagues had always been holding him back, discouraging his inbred interest in evolution and species. His new star status made him chafe for more independence and control, and he became more aggressive in his claims for material resources and authority. In so doing he ran afoul of the customs that subordinated individual interests to the collective good.

Money became a bone of contention early on. As Dobzhansky's

76. Provine, "Origins," p. 52; Dobzhansky and Queal, "Genic variation," p. 463; and Dobzhansky, "Experimental studies," pp. 345–46.

77. Dobzhansky to Stern, 16 Sept. 1937, CS.

78. Ibid.

field trips became more elaborate, they required more money. Not immense sums—his grant for the Alaska trip was just $450—but even modest sums were large in a culture that valued improvisation and parsimonious use of material resources. The fly group had mounted no collecting expedition since Sturtevant's in 1915, and Morgan was not accustomed to spending such sums. Dobzhansky became increasingly resentful of the boss's tight control of the Carnegie grant. For a time not even his ardent desire to go afield could overcome his fear of being turned down (in fact Morgan was sympathetic). The massive amount of wild material that Dobzhansky brought back from the field also entailed additional expense and strained communal facilities. As early as 1936 Dobzhansky had almost three hundred stocks under cultivation and was obliged to build and maintain his own stockroom, with special cooling equipment (*pseudoobscura* inhabited cooler places than its cosmopolitan cousins and did not thrive in Pasadena's desert climate). Dobzhansky also came to resent Morgan's one-man control of curricula and his efforts to build up the department of biochemistry at Caltech; the money, Dobzhansky felt, would have been much better spent on genetics.[79] Demerec encouraged his yearning to spread his wings: he had a right to expect adequate support for his work and would have to "cut loose from Morgan" to get it.[80]

Access to the group's human resources was a no less contentious issue. Working up masses of wild flies was immensely labor intensive, and Dobzhansky felt constantly in need of help. He wanted graduate students to pursue all the new leads that his system opened up. One of the attractions of the Texas job was the prospect of more students: "I have a flair for teaching beginning scientists," he wrote Demerec, "and here this is practically out of [the] question. There I am going to have more assistance and material possibilities than it will ever be possible here on account of Morgan's policy."[81] After 1937 Dobzhansky became more aggressive in enlisting students and postdoctoral fellows. This departure from the communal customs of the fly group was not appreciated by Dobzhansky's colleagues, or by students, who felt that he was treating them simply as hands for his own work. Tan and all three of Dobzhansky's graduate students soon transferred their allegiance to Sturtevant, who treated students not as "commodities for one's aggran-

79. Dobzhansky to Demerec, 7, 16 Jan. 1934, 16 Apr., 7, 20 May, 18, 25 Nov. 1936, MD; Dobzhansky to Patterson, 6 May 1936, TD; and Dobzhansky to Schultz, 5 Jan. 1939, JS.

80. Demerec to Dobzhansky, 4 Jan., 30 Mar. 1938, MD.

81. Dobzhansky to Demerec, 20 May 1936, MD.

dizement, but people who were interesting to have around." Jack Schultz, returning from Stockholm in 1939, was struck by the new mood of dissension and partisanship. This erosion of the old communal spirit, Schultz implied, was the result of Dobzhansky's heavy-handed efforts to recruit student workers and disciples.[82] Impelled by the unexpected productivity of *pseudoobscura,* Dobzhansky unwittingly violated the key customs of the group's moral economy.

Dobzhansky's relationship with Sturtevant continued to deteriorate. "Sturtevant suffers from bad dispositions," Dobzhansky reported to Demerec, "and [it] gets so that it is very difficult to talk to him nowadays. It is sad to watch him in this state."[83] Their rivalry became more open. Dobzhansky wanted to leave, but kept finding himself second in line to Sturtevant for the best jobs: first at Columbia, then at Harvard. Dobzhansky became convinced that Sturtevant did not mean to leave Caltech but just wanted to block Dobzhansky's chances to do so.[84] He later admitted that he had misjudged Sturtevant's motives, but such episodes left a bitter taste: "It is better simply not to care for a person than after many years to find out that he is not worth one's care."[85] As Dobzhansky grew more disenchanted with his colleagues, he tended more and more to retreat into a kind of internal exile, retiring in disgust from intellectual tiffs that once would have delighted him. "Why do these people create quarrels," he complained to Demerec, "why not work in peace, live and let live? It is detestable."[86]

When Dobzhansky was finally offered the Columbia post he accepted at once, not even waiting for Morgan to make a counteroffer. "It is no secret to you," he wrote Demerec, "that I am dead tired of [the] Pasadena environment, except for the natural environment—mountains, deserts and valleys, which I really love, and the loss of which

82. Schultz to Beadle, 31 July 1971, JS.

83. Dobzhansky to Demerec, 3 Mar. 1939, MD.

84. Dunn to Dobzhansky, 24 Apr. 1938, 2 May 1939, TD; and Dobzhansky to Demerec, 15 Nov., 24 Dec. 1939; F. L. Hisaw to Morgan, 27 Oct. 1939, Franz Schrader to Demerec, 29 June, 25 July 1935, and Demerec to Schrader, 12, 15, 17 July 1935, all in MD.

85. Dobzhansky to Demerec, 15 Nov. (quotation), 23 Nov. 1939, MD; and Demerec to Dobzhansky, 20 Nov. 1939, TD.

86. Dobzhansky to Demerec, 19 Oct. 1939 (quotation), also 24 Dec. 1938, 2 Feb., 4, 23, 27 Oct., 8, 15, 16 Nov. 1939, 30 Jan. 1940, MD; Demerec to Dobzhansky, 20 Nov. 1939, TD; Dobzhansky to Schultz, 5 Jan. 1939, JS; and Dobzhansky to Stern, 22 Aug. 1936, 16 Sept. 1937, CS.

I shall regret."[87] Sturtevant wrote him a touching note, apologizing for their disagreements and for all the "outrageous and inexcusable things" he had said to him over the years. "The place will seem strange without you, for none of the rest of us have your energy and go. Yes, you may be sure we'll miss you." It made Dobzhansky feel quite sentimental, but not reconciled.[88]

Conclusion

We have seen in earlier chapters how the construction of *Drosophila* in the 1910s made it highly efficient and productive for work on chromosomal mechanics, but less useful for work on the genetics of development and evolution. In this and the preceding chapter we saw how, in the 1930s, drosophilists got around these built-in limitations by inventing hybrid experimental systems, combining classical genetics with methods from other disciplines. I argued that it was this boundary crossing that brought about fundamental change. Beadle, Ephrussi, and Tatum imported transplantation and feeding from experimental embryology and nutritional chemistry. Dobzhansky crossed the boundary between lab and field and invented a mode of experiment with natural populations that combined field collecting and classical laboratory analysis in equal measure.

In both cases the standard fly, *D. melanogaster,* was eventually dislodged from its place of honor and replaced by other organisms: *D. pseudoobscura* and its wild cousins, and *Neurospora crassa.* The addition of new capabilities to the standard fly made possible new modes of practice and revealed the shortcomings of an instrument designed for chromosomal mechanics. *Neurospora* and *D. pseudoobscura* were not improvised, dual-purpose experimental technologies, as were mosaics and other trick systems in the 1920s, but specialized tools designed for new kinds of experimental work. Laboratories of developmental and evolutionary genetics were different from traditional fly labs, and in

87. Dobzhansky to Demerec, 30 Jan. (quotation), 3, 9 Feb. 1940, MD; Dunn to Dobzhansky, 1 Feb. 1940, TD; Dobzhansky to Dunn, 2 May 1938, 3 Mar. 1939, 30 Jan., 26 Feb. 1940, LCD; and Dobzhansky to Muller, 14 May 1940, HJM-A.

88. Sturtevant to Dobzhansky, 5 Feb. 1940, TD (quotation); Dobzhansky to Demerec, 9 Feb. 1940, MD; Dobzhansky to Schultz, 21 Mar. 1941, JS; Dobzhansky to Muller, 14 Apr. 1941, HJM-A; and Provine, "Origins," pp. 53, 56.

these new ecosystems the standard fly was no longer the favored species. It could not adapt and could not compete.

Dobzhansky's story reveals how the boundary between laboratory and field is a livelier and more interesting place than has been thought. Traditional histories of biology, which dwell on how experimentalists displaced natural historians from university departments, may have unwittingly created a false impression of the relations between lab and field biology. It is true, of course, that laboratory and field are different workplaces with different work cultures. But the boundary between them is neither immovable nor impermeable. Metz and Sturtevant brought wild flies indoors to be assimilated into a domestic, laboratory culture. Dobzhansky altered domestic laboratory practices to accommodate hoards of wild flies. Traditionally the field was an extension of the laboratory, a place to go occasionally. Dobzhansky's lab was, in contrast, an extension of the field. Similar stories could be told of the other places where drosophilists broke through the barriers separating lab and field: Berlin, Moscow, Austin. There, too, we may presume, new experimental practices were shaped by the particular, local ecologies of field and laboratory life.

The stories of Beadle and Dobzhansky illuminate the dynamic relation between experimental systems and the cultures and moral economies of experimental workplaces. The fly group's customs of easy assimilation and mobile cooperation were essential to the reinvention of developmental and evolutionary genetics. Just think of the many and varied ways in which Schultz and Sturtevant were crucial in the evolution of Beadle's and Dobzhansky's experimental practices: setting an example, inspiring bold gambits, collaborating, even perhaps assisting the divergence of new modes of practice through their disapproval. It is a striking fact, one of the most striking in the whole history of *Drosophila* genetics, that the developmental and evolutionary genetics of *Drosophila* both arose in the fly group. It was no accident: the fly group's productive and moral economies were designed to use diverse talents and to help the group stay in the vanguard.

Nor was it an accident that the fly group's boldest and most successful inventors of new systems—Beadle, Dobzhansky, Schultz—left to expand and develop their new practices in other places. Their experiences illustrate the point that really novel and productive modes of practice not only invigorate but also disrupt the productive and moral economy of working groups. Dobzhansky's genetics of natural populations competed with classical lines of work and threatened Sturtevant's investment in comparative genetics. Such conflict was inherent in the fly group's customs of retaining their best talents and encouraging them to do new things but at the same time imposing material and moral limits on overly expansive individual projects. Beadle escaped in time. Dobzhansky and Schultz stayed too long in a group in which practices that were too productive and too far from the mainstream were out of place (literally) and out of line. Beadle and Dobzhansky's systems outgrew drosophilists' traditional institutions and work culture in the 1940s and 1950s. Taken up by new generations of experimenters, these hybrid practices developed into disciplines that had their own special workplaces, material cultures, and ways of experimental life.

Appendix

TABLE A.1
Papers Published on *Drosophila* at Major Institutions, 1910–1937

Institution[a]	1910s					1920s					1930s				Total
	10/11	12/13	14/15	16/17	18/19	20/21	22/23	24/25	26/27	28/29	30/31	32/33	34/35	36/37	
United States															
Columbia/Caltech[b]	11	19	38	18	39	27	12	11	30	20	25	35	47	69	401
Cold Spring Harbor	1		2	8	2	7	9	2	7	3	3	8	14	12	78
Harvard	1	3		6	1	2								3	16
Indiana	3			1	4	3	2	2							15
Illinois			1	2	5	16	6	6	3	4	6	2	2	3	56
Rockefeller Inst			2	6		1		1	2	4	4	8	3		31
California				1			4	1	1	4				8	19
Texas					1	3	3	2	9	12	23	21	18	6	98
Amherst					1	1	2	2	2			2	2	13	25
Union Coll						2	6	5	1	2					16
Johns Hopkins Hyg						1	9	2	4	3		1			20
Michigan							1	2	2	2	3		1	1	12
Wooster Coll									3	2	3	3	3	1	15
Washington Univ										8	5	8	3		24
Western Reserve										1	3	3	4		11
New York Univ											5	11	14	15	45
Rochester												1	3	10	14
Columbia[b]												2	2	8	12
32 others	1	2	4	4	1	5	4	9	10	8	8	16	22	22	128
Total US	17	24	47	46	54	68	58	45	74	73	88	121	138	171	1024

Europe

Institution													Total
Paris/Geneva[c]				8	6		2	4	3	3	1	1	**23**
Oslo		1	1		4	3	7	2	3	3	1	1	**27**
Stockholm					1	1	3	1		1	1		**11**
KWI[d] Biology					1	3		9	13	12	10	11	**59**
KWI[d] Brain Res						6	6	12	14	9	27	16	**90**
Edinburgh								1		9	6	9	**25**
John Innes Inst									1	2	6	3	**12**
Ecole Normale										1	4	9	**14**
Univ Coll London											8	10	**18**
Rothschild Inst												5	**5**
33 others & unknown	3	4	2	1	4	11	6	7	22	12	25	21	**129**
Total Europe	**3**	**5**	**4**	**9**	**15**	**24**	**24**	**36**	**56**	**52**	**88**	**86**	**413**

Russia & Japan[e]

													Total
Total Russia				1		3	18	42	80	39	109	116	**408**
Total Japan						2	4	1	6	10	23	29	**75**

SOURCE: H. J. Muller, *Bibliography on the Genetics of Drosophila* (Edinburgh: Oliver & Boyd, 1939).

NOTE: Articles were scored according to where the research reported on was actually carried out. These data are usually supplied in American and most European journals.

The following types of publication were not counted: contributions to *DIS* and *Collecting Net*, popular articles and books, abstracts of papers published in full within three years, annual reports to the CIW, papers not specifically on *Drosophila* (Muller included some on salivary cytology, X-ray technology, and general theory), and strictly polemical articles. There is a subjective element in classifying some articles. My purpose was to construct a quantitative indicator of individuals' and institutions' output of experimental work.

a. Institutions are listed in chronological order by year of first published paper.

b. Papers by the Morgan group are combined under Columbia/Caltech. Papers from Columbia after 1928 are counted separately.

c. Emile Guyénot's group moved from the University of Paris to the University of Geneva.

d. KWI = Kaiser Wilhelm Institute

e. Articles by Russian and Japanese workers could not be systematically screened, hence only totals are given here. Some of their contributions may have been performed while they were visitors in U.S. or European institutions.

Appendix

Table A.2
Contributing Laboratories in the United States, 1910–1937, by Category of Institution

	Private univ.	Public univ.	Research inst.	Colleges	Med. schl./ hosp.	Museum/ society	Government	Total
1910/11	2	1	1		1			5
1912/13	2					2		4
1914/15	1	2	2			1		6
1916/17	3	4	2					9
1918/19	3	3	1	1				8
1920/21	4	4	2	2	1			13
1922/23	1	7	1	2	1	1		13
1924/25	3	8	2	3	1			17
1926/27	3	5	2	3	4			17
1928/29	5	6	2	2	2			17
1930/31	5	4	2	1	4			16
1932/33	11	5	2	2	4			24
1934/35	9	7	2	3	4			25
1936/37	13	6	1	3	2		1	26

SOURCE: Muller, *Bibliography on the Genetics of Drosophila*

NOTE: These figures give the number and type of laboratories active in *Drosophila* research per biennium, out of the 50 followed in table A.1.

Table A.3
Contributing Laboratories in Europe, 1910–1937, by Country

	Scandinavia, Netherlands	France, Switzerland	Germany, Austria, Czech.	Great Britain	Other	Total Europe
1910/11	1	1				2
1912/13		2	1			3
1914/15		1		1		2
1916/17	1					1
1918/19	1		1	1		3
1920/21	1	1				2
1922/23	3	1	1	1		6
1924/25	3	1	2	1		7
1926/27	2	1	4			7
1928/29	3	1	7	1		12
1930/31	2	1	8	2	3	16
1932/33	3	2	4	4	2	15
1934/35	2	2	10	3	3	20
1936/37	2	3	8	5	2	20

SOURCE: Muller, *Bibliography on the Genetics of Drosophila*

NOTE: These figures give the number and location of laboratories active in *Drosophila* research per biennium, out of the 43 followed in table A.1.

Abbreviations

AHS	Alfred H. Sturtevant Papers, Archives, California Institute of Technology, Pasadena
APS	American Philosophical Society Library, Philadelphia
CBD	Charles B. Davenport Papers, APS
CIW	Carnegie Institution of Washington, Washington, D.C.
CS	Curt Stern Papers, APS
CZ	Charles Zeleny Papers, Archives, University of Illinois, Urbana
DIS	*Drosophila Information Service*
DSB	*Dictionary of Scientific Biography* (New York: Scribners, 1970–80), and Supplement II (1990)
FP	Fernandus Payne Papers, Lilly Library, University of Indiana, Bloomington
GEB	General Education Board Papers, Rockefeller Archive Center, North Tarrytown, New York
GWB	George W. Beadle Papers, Archives, California Institute of Technology, Pasadena
HJM-A	Hermann J. Muller Papers, alphabetical series, Lilly Library, Indiana University, Bloomington
HJM-C	Hermann J. Muller Papers, chronological series, Lilly Library, Indiana University, Bloomington
JS	Jack Schultz Papers, APS
LCD	Leslie C. Dunn Papers, APS
MD	Milislav Demerec Papers, APS
OM	Otto Mohr Papers, Oslo
RAM	Robert A. Millikan Papers, Archives, California Institute of Technology, Pasadena
RF	Rockefeller Foundation Papers, Rockefeller Archive Center, North Tarrytown, New York
RG	Richard Goldschmidt Papers, Bancroft Library, University of California, Berkeley

SW   Sewell Wright Papers, APS
TD   Theodosius Dobzhansky Papers, APS
THM-APS   Thomas Hunt Morgan Papers, APS
THM-CIT   Thomas Hunt Morgan Papers, Archives, California Institute of
            Technology, Pasadena
UCDG   University of California Department of Genetics Papers, APS
WP   William Provine Correspondence, courtesy of William Provine,
            Ithaca, New York

Citations to collections may include box number, box and folder numbers (e.g., 3.20), or box number and file name. Citations to GEB and RF include record group number (e.g., 1.1) and series, box, and folder number (e.g., 205 3.19). Documents not so identified are filed by name of correspondent.

Bibliography

The bibliography is organized in three parts: secondary works in general history and sociology of science, including some histories of biology focused on practice; histories of biology, including histories and reminiscences by historical figures; and selected scientific works that are cited frequently or are of general interest. Works of a given author are listed chronologically.

History and Sociology of Experimental Practice

Bijker, Wiebe E., Thomas P. Hughes, and Trevor Pinch, editors. *The Social Construction of Technological Systems*. Cambridge: MIT Press, 1987.

Budiansky, Stephen. *The Covenant of the Wild: Why Animals Chose Domestication*. New York: William Morrow, 1992.

Clarke, Adele E. "Research materials and reproductive science in the United States, 1910–1940." In Gerald Geison, editor, *Physiology in the American Context, 1850–1940*, 323–50. Bethesda, Md.: American Physiological Society, 1987.

Clarke, Adele E., and Joan Fujumura, editors. *The Right Tools for the Job: At Work in Twentieth-Century Life Sciences*. Princeton: Princeton University Press, 1992.

Clarke, Adele E., and Joan Fujimura. "What tools? Which jobs? Why right?" In Clarke and Fujimura, eds., *Right Tools for the Job*, 3–44 (excellent bibliography).

Clarke, Adele E., and Elihu Gerson. "Symbolic interactionism in science studies." In Howard S. Becker and Michael McCall, editors, *Symbolic Interactionism and Cultural Studies*, 170–214. Chicago: University of Chicago Press, 1990.

Clause, Bonnie T. "The Wistar rat as a right choice: Establishing mammalian standards and the ideal of a standardized mammal." *Journal of the History of Biology* 26 (1993): 329–49.

Coleman, William, and Frederic L. Holmes, editors. *The Investigative Enterprise: Experimental Physiology in Nineteenth-Century Medicine*. Berkeley: University of California Press, 1988.

Collins, Harry M. *Changing Order: Replication and Induction in Scientific Practice*. Second edition. Chicago: University of Chicago Press, 1992.

Cronon, William. *Changes in the Land: Indians, Colonists, and the Ecology of New England.* New York: Hill & Wang, 1983.

———. "Modes of prophecy and production: Placing nature in history." *Journal of American History* 76 (1990): 1122–31.

———. *Nature's Metropolis: Chicago and the Great West.* New York: Norton, 1991.

———. "A place for stories: Nature, history, and narrative." *Journal of American History* 78 (1992): 1347–76.

Elzen, Boelie. "Two ultracentrifuges: A comparative study of the social construction of artifacts." *Social Studies of Science* 16 (1986): 621–62.

Falconer, Isobel. "J. J. Thomson's work on positive rays, 1906–1914." *Historical Studies in the Physical and Biological Sciences* 18 (1988): 267–310.

Frank, Robert G., Jr. "The telltale heart: Physiological instruments, graphic methods, and clinical hopes, 1854–1914." In Coleman and Holmes, eds., *Investigative Enterprise*, 211–90.

Fujimura, Joan. "Constructing doable problems in cancer research: Articulating alignment." *Social Studies of Science* 17 (1987): 257–93.

———. "The molecular biology bandwagon in cancer research: Where social worlds meet," *Social Problems* 35 (1988): 261–83.

Galison, Peter. "Bubble chambers and the experimental workplace." In Peter Achinstein and Owen Hannaway, editors, *Observation, Experiment, and Hypothesis in Modern Physical Science,* 309–73. Cambridge: MIT Press, 1985.

———. *How Experiments End.* Chicago: University of Chicago Press, 1987.

Galison, Peter, and Alexi Assmus. "Artificial clouds, real particles." In Gooding, Pinch, Schaffer, eds., *Uses of Experiments,* 225–74.

Gooding, David. "'In nature's school': Faraday as an experimentalist." In David Gooding and Frank A. J. L. James, editors, *Faraday Rediscovered,* 105–35. London: Macmillan, 1985.

Gooding, David, Trevor Pinch, and Simon Schaffer, editors. *The Uses of Experiment.* Cambridge: Cambridge University Press, 1989.

Hagen, Joel B. "Experimentalists and naturalists in twentieth-century botany: Experimental taxonomy." *Journal of the History of Biology* 17 (1984): 249–70.

Hamlin, Christopher. "Robert Warington and the moral economy of the aquarium." *Journal of the History of Biology* 19 (1986): 131–53.

Hannaway, Owen. "Laboratory design and the aim of science: Andreas Libavius versus Tycho Brahe." *Isis* 77 (1986): 585–610.

Haraway, Donna. *Primate Visions: Gender, Race, and Nature in the World of Modern Science.* New York: Routledge, 1989.

Howell, Joel D. "Early perceptions of the electrocardiogram: From arrythmia to infarction." *Bulletin of the History of Medicine* 58 (1984): 83–98.

Hughes, Thomas P. "The evolution of large technological systems." In Bijker, Hughes, and Pinch, eds., *Social Construction,* 51–82.

———. *Networks of Power: Electrification in Western Society, 1880–1930.* Baltimore: Johns Hopkins University Press, 1983. Chapter 1.

Hunt, Bruce. "Experimenting on the ether: Oliver Lodge and the great whirling machine." *Historical Studies in the Physical and Biological Sciences* 16 (1986): 111–34.

Kidder, Tracy. *The Soul of a New Machine.* Boston: Little, Brown, 1981.

Knorr-Cetina, Karen. "Tinkering toward success: Prelude to a theory of scientific practice." *Theory and Society* 8 (1979): 347–76.

Knorr-Cetina, Karen, R. Krohn, and R. Whitley, editors. *The Social Process of Scientific Investigation.* Dordrecht: Reidel, 1984.

Kohler, Robert E. "The Ph.D. machine: Building on the collegiate base." *Isis* 81 (1990): 638–62.

———. "Systems of production: *Drosophila, Neurospora,* and biochemical genetics." *Historical Studies in the Physical and Biological Sciences* 22 (1991): 87–130.

———. "Drosophila and evolutionary genetics: The moral economy of scientific practice," *History of Science* 29 (1991): 335–75.

———. *Partners in Science: Foundations and Natural Scientists, 1900–1945.* Chicago: University of Chicago Press, 1991.

———. "Drosophila: A life in the lab." *Journal of the History of Biology* 26 (1993): 281–310.

Kuklick, Henrika. *The Savage Within: The Social History of British Anthropology, 1885–1945.* Cambridge: Cambridge University Press, 1992.

Lankford, John. "Amateurs versus professionals: The controversy over telescope size in late Victorian science." *Isis* 72 (1981): 11–28.

Latour, Bruno. "Give me a laboratory and I will raise the world." In Karin Knorr-Cetina and Michael Mulkay, editors, *Science Observed,* 141–70. London: Sage, 1983.

———. *Science in Action.* Cambridge: Harvard University Press, 1987.

———. *The Pasteurization of France.* Cambridge: Harvard University Press, 1988.

Latour, Bruno, and Steve Woolgar. *Laboratory Life: The Social Construction of Scientific Facts.* Beverley Hills, Calif.: Sage, 1979.

Lederer, Susan E. "Political animals: The shaping of biomedical research literature in twentieth-century America." *Isis* 83 (1992): 61–79.

Le Grand, Homer, editor. *Experimental Inquiries: Historical, Philosophical, and Social Studies of Experimentation in Science.* Dordrecht: Kluwer Academic, 1990.

Lenoir, Timothy. "Models and instruments in the development of electrophysiology, 1845–1912." *Historical Studies in the Physical and Biological Sciences* 17 (1987): 1–54.

Lynch, Michael. *Art and Artifact in Laboratory Science.* London: Routledge & Kegan Paul, 1985.

———. "Sacrifice and the transformation of the animal body into a scientific object: Laboratory culture and ritual practice in the neurosciences." *Social Studies of Science* 18 (1988): 265–89.

Miller, David P. "Values redivivus?" *Social Studies of Science* 22 (1992): 419–27.

Mitman, Greg, and Anne Fausto-Sterling. "Whatever happened to *Planaria?* C. M. Child and the physiology of inheritance." In Clarke and Fujimura, eds., *Right Tools for the Job,* 172–96.

Ophir, Adi, and Steven Shapin. "The place of knowledge: A methodological survey." *Science in Context* 4 (1991): 3–21.

Pang, Alex Soojung-Kim. "Spheres of Interest: Imperialism, Culture, and Practice in British Socal Eclipse Expeditions, 1860–1914." Ph.D. diss., University of Pennsylvania, 1991.

———. "The social event of the season: Eclipse expeditions and Victorian culture." *Isis* 83 (1993): 252–77.

Pickering, Andrew. *Constructing Quarks: A Sociological History of Particle Physics.* Chicago: University of Chicago Press, 1984.

———, editor. *Science as Practice and Culture.* Chicago: University of Chicago Press, 1992.

Pinch, Trevor. *Confronting Nature: The Sociology of Solar Neutrino Detection.* Dordrecht: Reidel, 1986.

Rindos, David. *The Origins of Agriculture: An Evolutionary Approach.* New York: Academic Press, 1984.

Ritvo, Harriet. *The Animal Estate: The English and Other Creatures in the Victorian Age.* Cambridge: Harvard University Press, 1987.

Sauer, Carl. "Man's dominance by use of fire." In *Selected Essays, 1963–1975,* 129–56. Berkeley: Turtle Island Foundation, 1981.

Schaffer, Simon. "Scientific discoveries and the end of natural philosophy." *Social Studies of Science* 16 (1986): 387–420.

Scott, James C. *The Moral Economy of the Peasant: Rebellion and Subsistence in Southeast Asia.* New Haven: Yale University Press, 1976.

Secord, James A. "Extraordinary experiment: Electricity and the creation of life in Victorian England." In Coleman and Holmes, eds., *Investigative Enterprise,* 357–84.

Serpell, James. *In the Company of Animals.* Oxford: Basil Blackwell, 1986.

Shapin, Steven. "Pump and circumstance: Robert Boyle's literary technology." *Social Studies of Science* 14 (1984): 481–520.

———. "The house of experiment in seventeenth-century England." *Isis,* 79 (1988): 373–404.

———. "The invisible technician." *American Scientist* 77 (1989): 554–63.

———. "Who was Robert Hooke?" In Michael Hunter and Simon Schaffer, editors, *Robert Hooke: New Studies*, 253–86. Woodbridge, U.K.: Boydell Press, 1989.

———. "'A scholar and a gentleman': The problematic identity of the scientific practitioner in early modern England." *History of Science* 29 (1991): 279–327.

Shapin, Steven, and Simon Schaffer. *Leviathan and the Air Pump: Hobbes, Boyle, and the Experimental Life*. Princeton: Princeton University Press, 1985.

Silver, Timothy. *A New Face on the Countryside: Indians, Colonists, and Slaves in South Atlantic Forests, 1500–1800*. Cambridge: Cambridge University Press, 1990.

Star, Susan L. *Regions of the Mind: Brain Research and the Quest for Scientific Certainty*. Palo Alto: Stanford University Press, 1989.

Star, Susan L., and James Griesemer. "Institutional ecology, 'translations,' and boundary objects: Amateurs and professionals in Berkeley's Museum of Vertebrate Zoology." *Social Studies of Science* 19 (1989): 387–420.

Thompson, Edward P. "The moral economy of the English crowd in the eighteenth century." *Past and Present* 50 (1971): 76–136. Reprinted in Thompson, *Customs in Common*, 185–258. New York: New Press, 1991.

———. "The moral economy reviewed." In Thompson, *Customs in Common*, 259–351.

Tobey, Ronald C. *Saving the Prairies: The Life Cycle of the Founding School of American Plant Ecology, 1895–1955*. Berkeley: University of California Press, 1981.

Traweek, Sharon. *Beamtimes and Lifetimes: The World of High Energy Physics*. Cambridge: Harvard University Press, 1988.

Winner, Langdon. "Do artifacts have politics?" *Daedalus* 109 (1980): 121–36.

Worster, Donald. "Transformations of the earth: Toward an agroecological perspective in history," *Journal of American History* 76 (1990): 1087–1106.

———. "Seeing beyond culture." *Journal of American History* 76 (1990): 1142–47.

Wright, James R., Jr. "The development of the frozen section technique, the evolution of surgical biopsy, and the origins of surgical pathology." *Bulletin of the History of Medicine* 59 (1985): 295–326.

History of Biology

Adams, Mark B. "The founding of population genetics: Contributions of the Chetverikov school, 1924–1934." *Journal of the History of Biology* 1 (1968): 23–39.

———. "Toward a synthesis: Population concepts in Russian evolutionary thought, 1925–1935." *Journal of the History of Biology* 3 (1970): 107–29.

———. "Sergei Chetverikov, the Kol'tsov Institute, and the evolutionary synthesis." In Mayr and Provine, eds., *Evolutionary Synthesis*, 242–78.

Allen, Garland E. "Thomas Hunt Morgan and the problem of natural selection." *Journal of the History of Biology* 1 (1968): 113–39.

———. "Hugo de Vries and the reception of the 'mutation theory.'" *Journal of the History of Biology* 2 (1969): 55–87.

———. "T. H. Morgan and the emergence of a new American biology." *Quarterly Review of Biology* 44 (1969): 168–88.

———. "Naturalists and experimentalists: The genotype and the phenotype." *Studies in the History of Biology* 3 (1970): 179–209.

———. "The introduction of *Drosophila* into the study of heredity and evolution, 1900–1910." *Isis* 66 (1975): 322–33.

———. *Thomas Hunt Morgan: The Man and His Science.* Princeton: Princeton University Press, 1978.

———. "T. H. Morgan and the split between embryology and genetics, 1910–35." In Horder et al., eds., *History of Embryology*, 113–46.

Anderson, Thomas F. "Jack Schultz." *Biographical Memoirs of the National Academy of Sciences* 47 (1975): 393–427.

Beadle, George W. "Chemical genetics." In Leslie C. Dunn, editor, *Genetics in the Twentieth Century*, 221–39. New York: Macmillan, 1951.

———. Foreword to *Genetics and Evolution: Selected Papers of A. H. Sturtevant*, ed. E. B. Lewis. San Francisco: Freeman, 1961, iii–iv.

———. "Biochemical genetics: Some recollections." In John Cairns, Gunther S. Stent, and James D. Watson, editors, *Phage and the Origins of Molecular Biology*, 23–32. Cold Spring Harbor, N.Y.: Cold Spring Harbor Laboratory, 1966.

———. "Alfred Henry Sturtevant (1891–1970)." *American Philosophical Society Yearbook* (1970): 166–71.

———. "Recollections." *Annual Review of Biochemistry* 43 (1974): 1–13.

Benson, Keith R., Jane Maienschein, and Ronald Rainger, editors. *The Expansion of American Biology.* New Brunswick, N.J.: Rutgers University Press, 1991.

Bowler, Peter J. "Hugo de Vries and Thomas Hunt Morgan: The mutation theory and the spirit of Darwinism." *Annals of Science* 35 (1978): 55–73.

Burian, Richard M., Jean Gayon, and Doris Zallen. "The singular fate of genetics in the history of French biology, 1900–1940." *Journal of the History of Biology* 21 (1988): 357–402.

Carlson, Elof A. *The Gene: A Critical History.* Philadelphia: Saunders, 1966.

————. "The *Drosophila* group: The transition from the Mendelian unit to the individual gene." *Journal of the History of Biology* 7 (1974): 31–48.

————. *Genes, Radiation, and Society: The Life and Work of H. J. Muller.* Ithaca: Cornell University Press, 1981.

Ephrussi, Boris. "The cytoplasm and somatic cell variation." *Journal of Cellular and Comparative Physiology* 52 supplement 1 (1958): 35–53.

Dobzhansky, Theodosius. Oral history. 1962. Butler Library, Columbia University, New York.

————. *The Roving Naturalist: The Travel Letters of Theodosius Dobzhansky,* ed. Bentley Glass. Philadelphia: American Philosophical Society, 1980.

————. *Dobzhansky's Genetics of Natural Populations, I–XLIII.* Edited by Richard C. Lewontin, John A. Moore, William B. Provine, and Bruce Wallace. New York: Columbia University Press, 1981.

Gilbert, Scott F. "The embryological origins of the gene theory." *Journal of the History of Biology* 11 (1978): 307–51.

Glass, Bentley. "Milislav Demerec." *Biographical Memoirs of the National Academy of Sciences* 42 (1971): 1–27.

Harwood, Jonathan. "The reception of Morgan's chromosomal theory in Germany: Inter-war debate over cytoplasmic inheritance." *Medizinhistorisches Journal* 19 (1984): 3–32.

————. "Geneticists and the evolutionary synthesis in interwar Germany." *Annals of Science* 42 (1985): 279–301.

————. "National styles in science: Genetics in Germany and the United States between the world wars." *Isis* 78 (1987): 390–414.

————. *Styles of Scientific Thought: The German Genetics Community, 1900–1933.* New York: Oxford University Press, 1992.

Horder, T. J., J. A. Witkowski, and C. C. Wylie, editors. *A History of Embryology.* Cambridge: Cambridge University Press, 1986.

Horowitz, Norman H. "George Wells Beadle." *Biographical Memoirs of the National Academy of Sciences* 59 (1990): 27–52.

————. "Fifty years ago: The *Neurospora* revolution." *Genetics* 127 (1991): 631–35.

Kay, Lily E. "Selling pure science in wartime: The biochemical genetics of G. W. Beadle." *Journal of the History of Biology* 22 (1989): 73–101.

————. *The Molecular Vision of Life: Caltech, the Rockefeller Foundation, and the Rise of the New Biology.* New York: Oxford University Press, 1993.

Kevles, Daniel J. "Genetics in the United States and Great Britain, 1890–1930: A review with speculations." *Isis* 71 (1980): 441–55.

Kimmelman, Barbara A. "Organisms and interests in scientific research: R. A. Emerson's claims for the unique contributions of agricultural genetics." In Clarke and Fujimura, eds., *Right Tools for the Job,* 198–232.

Bibliography

Krementsov, Nikolai. "Dobzhansky and Russian entomology: The origin of his ideas on species and speciation." In Mark Adams, editor, *The Evolution of Theodosius Dobzhansky*. Princeton: Princeton University Press, 1994.

Lederberg, Joshua. "Edward Lawrie Tatum." *Biographical Memoirs of the National Academy of Sciences* 59 (1990): 357–86.

Lewis, Edward B. "Alfred H. Sturtevant." *DSB* 13: 133–38.

Lewontin, Richard C. "The scientific work of Th. Dobzhansky." In *Dobzhansky's Genetics*, ed. Lewontin et al., 93–115.

Lwoff, André. "Recollections of Boris Ephrussi." *Somatic Cell Genetics* 5 (1979): 677–79.

Maienschein, Jane. Introduction of Maienschein, editor, *Defining Biology: Lectures from the 1890s*, 3–50. Cambridge: Harvard University Press, 1986.

———. *Transforming Traditions in American Biology, 1880–1915*. Baltimore: Johns Hopkins University Press, 1991.

Mayr, Ernst, and William B. Provine, editors. *The Evolutionary Synthesis: Perspectives on the Unification of Biology*. Cambridge: Harvard University Press, 1980.

Morgan, Thomas H. "Edmund Beecher Wilson 1856–1939." *Biographical Memoirs of the National Academy of Sciences* 21 (1940): 315–42.

———. "Calvin Blackman Bridges." *Biographical Memoirs of the National Academy of Sciences*, 22 (1941), 31–48. A longer unpublished draft of this biographical memoir is in THM-APS.

———. "Genesis of the white-eyed mutant." *Journal of Heredity* 33 (1942): 91–92.

The Naples Zoological Station and the Marine Biological Laboratory: One Hundred Years of Biology. *Biological Bulletin* 168 supplement (1985): 1–207.

James V. Neel. "Curt Stern." *Biographical Memoirs of the National Academy of Sciences* 56 (1987): 443–73.

Paul, Diane B., and Barbara A. Kimmelman. "Mendel in America: Theory and practice, 1900–1919." In Rainger et al., eds., *American Development of Biology*, 281–310.

Paul, Diane B., and Costas B. Krimbas. "Nikolai V. Timoféeff-Ressovsky." *Scientific American* (Feb. 1992): 86–92.

Pauly, Philip. "Summer resort and scientific discipline: Woods Hole and the structure of American biology, 1882–1925." In Rainger et al., eds., *American Development of Biology*, 121–50.

———. "The development of high school biology: New York City, 1900–1925." *Isis* 82 (1991): 662–88.

Provine, William B. *The Origins of Theoretical Population Genetics*. Chicago: University of Chicago Press, 1971.

———. "The role of mathematical population geneticists in the evolutionary synthesis of the 1930s and 1940s." *Studies in the History of Biology* 2 (1978): 167–92.

———. "Origins of the genetics of natural populations series." In *Dobzhansky's Genetics,* ed. Lewontin et al., 5–79.

———. *Sewall Wright and Evolutionary Biology.* Chicago: University of Chicago Press, 1986.

Rainger, Ronald, Keith R. Benson, and Jane Maienschein, editors. *The American Development of Biology.* Philadelphia: University of Pennsylvania Press, 1988.

Ravin, Arnold W. "Francis J. Ryan (1916–1963)." *Genetics* 84 (1976): 1–25.

Roll-Hansen, Nils. "*Drosophila* genetics: A reductionist research program." *Journal of the History of Biology* 11 (1978): 159–210.

Roman, Herschel. "Boris Ephrussi." *Annual Review of Genetics* 14 (1980): 447–50.

Rosenberg, Charles E. "Factors in the development of genetics in the United States: Some suggestions." *Journal of the History of Medicine* 22 (1967): 27–46.

Sander, Klaus. "The role of genes in ontogenesis—Evolving concepts from 1883 to 1983 as perceived by an insect embryologist." In Horder et al., eds., *History of Embryology,* 363–95.

Sapp, Jan. "The struggle for authority in the field of heredity, 1900–1932: New perspectives on the rise of genetics." *Journal of the History of Biology* 16 (1983): 311–42.

———. "Inside the cell: Genetic methodology and the case of the cytoplasm." In J. A. Schuster and R. R. Yeo, editors, *The Politics and Rhetoric of Scientific Method,* 167–202. Dordrecht: Reidel, 1986.

———. *Beyond the Gene: Cytoplasmic Inheritance and the Struggle for Authority in Genetics.* New York: Oxford University Press, 1987.

Schultz, Jack. "Innovators and controversies." *Science* 157 (1967): 296–301.

Secord, James A. "Nature's fancy: Charles Darwin and the breeding of pigeons." *Isis* 72 (1981): 163–86.

———. "Darwin and the breeders: A social history." In David Krohn, editor, *The Darwinian Heritage,* 519–42. Princeton: Princeton University Press, 1986.

Sturtevant, Alfred H. "Thomas Hunt Morgan." *Biographical Memoirs of the National Academy of Sciences* 33 (1959): 283–325. Other published reminiscences are in AHS.

———. *A History of Genetics.* New York: Harper & Row, 1965.

Wald, George. "Selig Hecht (1892–1947)." *Journal of General Physiology* 32 (1948): 1–16.

Bibliography

Drosophila Genetics: Selected Primary Sources

Beadle, George W. "Genetics and metabolism in *Neurospora*," *Physiological Reviews* 25 (1945): 643–63.

Beadle, George W., and Boris Ephrussi. "The differentiation of eye pigments in *Drosophila* as studied by transplantation." *Genetics* 21 (1936): 225–47.

Beadle, George W., and Edward L. Tatum. "Experimental control of development and differentiation." *American Naturalist* 75 (1941): 107–16.

Bridges, Calvin B. "Gametic and observed ratios in *Drosophila*." *American Naturalist* 55 (1921): 51–61.

———. "Apparatus and methods for *Drosophila* culture." *American Naturalist* 66 (1932): 250–73.

Bridges, Calvin B., and H. H. Darby. "Culture media for *Drosophila* and the pH of media." *American Naturalist* 67 (1933): 437–72.

———. "A system of temperature control." *Journal of the Franklin Institute* 215 (1933): 723–30.

Bridges, Calvin B., and T. H. Morgan. "The Second-Chromosome Group of Mutant Characters." In *Contributions to the Genetics of Drosophila melanogaster,* Carnegie Institution of Washington publication no. 278. Washington, 1919, 125–342.

———. *The Third-Chromosome Group of Mutant Characters of Drosophila melanogaster.* Carnegie Institution of Washington publication no. 327. Washington, 1923.

Bridges, Calvin B., and T. M. Olbrycht. "The multiple stock 'Xple' and its use." *Genetics* 11 (1926): 41–56.

Demerec, Milislav, editor. *Biology of Drosophila.* New York: Wiley, 1950.

Dobzhansky, Theodosius, and Alfred H. Sturtevant. "Inversions in the chromosomes of *Drosophila psudoobscura*." *Genetics* 23 (1938): 28–64.

Drosophila Information Service 1 (1934) to date.,

Ephrussi, Boris. "Chemistry of 'eye color hormones' of *Drosophila*." *Quarterly Review of Biology* 17 (1942): 327–38.

Ephrussi, Boris, and George W. Beadle. "A technique of transplantation for *Drosophila*." *American Naturalist* 70 (1936): 218–25.

Haecker, Valentin. *Entwicklungsgeschichtliche Eigenschaftsanalyse (Phaenogenetik).* Stuttgart: Fischer, 1918.

———. "Phänogenetisch gerichtete Bestrebungen in Amerika." *Zeitschrift für induktive Abstammungs- und Vererbungslehre* 41 (1926): 232–38.

Harrison, Ross G. "Embryology and its relations." *Science* 85 (1937): 369–74.

Herskowitz, Irwin H. *Bibliography on the Genetics of Drosophila*, Part 2. Edinburgh: Oliver & Boyd, [c. 1952].

————. *Bibliography on the Genetics of Drosophila,* Part 3. Bloomington: Indiana University Press, 1958.

Lutz, Frank B. "The merits of the fruit fly." *School Science and Mathematics* 7 (1907): 672–73.

Morgan, Thomas H. *Evolution and Adaptation.* New York: Macmillan, 1903.

————. "The origin of species through selection contrasted with their origin through the appearance of definite variations." *Popular Science Monthly* 67 (1905): 54–65.

————. "For Darwin." *Popular Science Monthly* 74 (1909): 367–80.

————. "Factors and unit characters in Mendelian heredity." *American Naturalist* 47 (1913): 5–16.

————."The mechanism of heredity as indicated by the inheritance of linked characters." *Popular Science Monthly* 84 (1914): 5-16.

————. "The theory of the gene." *American Naturalist* 51 (1917): 513–44.

————. "Genetics and the physiology of development." *American Naturalist* 60 (1926): 489–515.

————. *Embryology and Genetics.* New York: Columbia University Press, 1934.

Morgan, Thomas H., and Calvin B. Bridges. "The origin of gynandromorphs." In Morgan and Bridges, editors, *Contributions to the Genetics of Drosophila melanogaster.* Carnegie Institution of Washington publication no. 278. Washington, 1919, 1–122.

Morgan, Thomas H., Calvin B. Bridges, and Alfred H. Sturtevant. "The Genetics of *Drosophila.*" *Bibliographia Genetica* 2 (1925): 1–262.

Morgan, Thomas H., Alfred H. Sturtevant, Hermann J. Muller, and Calvin B. Bridges. *The Mechanism of Mendelian Heredity.* New York: Holt, 1915.

Muller, Hermann J. *Bibliography on the Genetics of Drosophila.* Edinburgh: Oliver & Boyd, 1939.

Patterson, John T., and Wilson S. Stone. *Evolution in the Genus Drosophila.* New York: Macmillan, 1952.

Ryan, Frances J., George W. Beadle, and Edward I. Tatum. "The tube method of measuring the growth rate of neurospora." *American Journal of Botany* 30 (1943): 784–99.

Schultz, Jack. "Aspects of the relation between genes and development in *Drosophila.*" *American Naturalist* 69 (1935): 30–54.

Spencer, Warren P. "Collection and laboratory culture." In Demerec, *Biology of Drosophila,* 535–87.

Sturtevant, Alfred H. "Notes on North American Drosophilidae, with descriptions of twenty-three new species." *Annals of the Entomological Society of America* 9 (1916): 323–43.

313

————. *Contributions to the Genetics of Drosophila simulans and Drosophila melanogaster.* Carnegie Institution of Washington publication no. 399. Washington, 1929.

————. "On the subdivision of the genus *Drosophila.*" *Proceedings of the National Academy of Sciences* 25 (1939): 337–53.

————. "Physiological aspects of genetics." *Annual Review of Physiology* 34 (1941): 41–56.

————. *The Classification of the Genus Drosophila, with Description of Nine New Species.* University of Texas Publication no. 4213. Austin, 1942.

Tatum, Edward L., and George W. Beadle. "The relation of genetics to growth-factors and hormones." *Growth* 6 supplement (1942): 27–37.

Index

Agol, Israel, 161
Altenburg, Edgar, 62, 92, 93, 140
Anderson, Ernest G., 82, 92, 125–26, 212
Animals, domesticated, 9–10, 28–30. *See also Drosophila,* domesticated

Barrows, W. M., 144
Beadle, George W., 82, 127, 144, 211–12, 253; and Boris Ephrussi, 208–9, 214–27, 227–32; and *Neurospora,* 233–44, 248; as transdisciplinary, 177, 189, 247–48, 291–92; transfusion method of, 227–37; transplantation method of, 205, 209–10, 217–24
Belar, Karl, 256
Biochemical genetics, 233–42. *See also* Genetics and biochemistry
Bishop, Maydelle, 147
Boche, Robert D., 260
Bonnier, Gert, 145–46
Bridges, Calvin B., 36, 96, 139; and *Drosophila* exchange network, 129, 130, 134, 142–43, 145–46, 151–53, 156, 159; and *Drosophila Information Service,* 157, 162–67; and mapping project, 61–65, 156; and mapping technology, 65–71, 75–78, 84–87; and T. H. Morgan, 115, 120, 123–24, 126–27; in Morgan group, 92, 98, 99–105, 107–10, 113–17, 253, 256, 268

Bush, Vannevar, 129, 131
Butenandt, Adolf, 232, 234

California Institute of Technology, 122, 124, 129, 159, 256, 275, 278, 289. *See also* Morgan group (Caltech)
Calkins, Gary N., 117–18
Carnegie Institution of Washington, 280, 282; as "employer", 92, 115, 122, 123, 124, 130; and *Drosophila* supply, 130, 158–59; and grant to Morgan group, 106–10, 121, 129, 131, 173, 178, 275, 289. *See also* Cold Spring Harbor Laboratory
Carver, Gail, 36, 93
Caspersson, Torbjörn, 129, 193
Castle, William E., 23, 27, 30, 34–35, 39, 41, 46
Cattell, Eleth, 36
Chen, T. Y., 244
Chetverikov, Sergei, 251–52, 257, 264, 270
Child, George P., 182
Clausen, Jens, 282
Clausen, Roy, 111, 139
Coccinella, 254, 261
Cold Spring Harbor Laboratory, 23–24, 31–33, 34, 177–78, 199, 172, 275; "dust-up" (1919), 93, 108–109, 121, 201; as rival center, 138–39, 151, 156, 158–60, 164, 165. *See also* Metz, Charles
Columbia University, 24, 36, 38, 92;

DEMCO